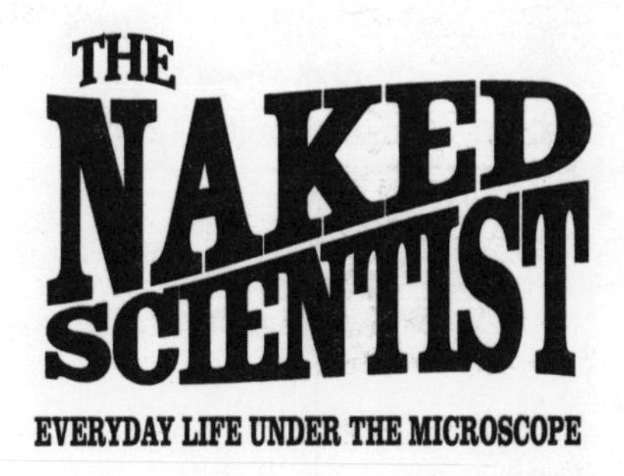
THE
NAKED
SCIENTIST
EVERYDAY LIFE UNDER THE MICROSCOPE

Also by Chris Smith

The Naked Scientist:
The Science of Everyday Life Laid Bare

Crisp Packet Fireworks:
Maverick Science to Try at Home
(with Dave Ansell)

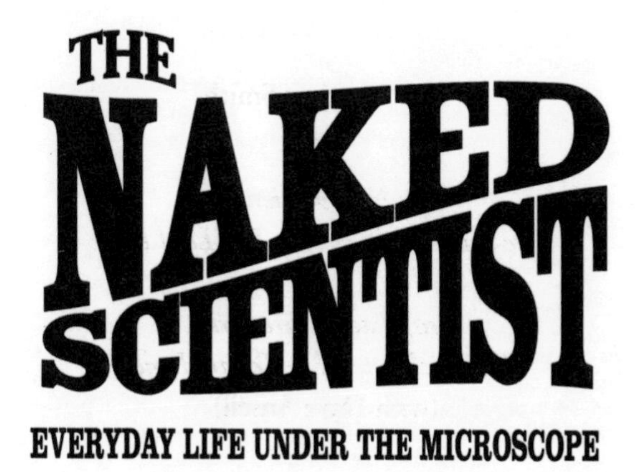

EVERYDAY LIFE UNDER THE MICROSCOPE

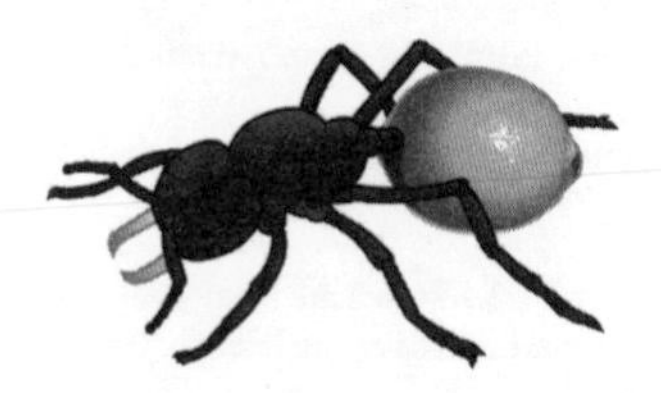

CHRIS SMITH

Little, Brown

LITTLE, BROWN

First published in Great Britain in 2012 by Little, Brown
First published in Australia in 2008 by William Heinemann,
an imprint of Random House Australia

A CIP catalogue record for this book
is available from the British Library.

ISBN 978-1-4087-0380-9

Printed and bound in Great Britain by
Clays Ltd, St Ives plc

Papers used by Little, Brown are from well-managed forests
and other responsible sources.

Little, Brown
An imprint of
Little, Brown Book Group
100 Victoria Embankment
London EC4Y 0DY

An Hachette UK Company
www.hachette.co.uk

www.littlebrown.co.uk

This book is dedicated to my wonderful family
– Sarah, Amelia and Tim – All my love.

INTRODUCTION

Who says arguments aren't constructive? I say that because this book is largely the result of a row with my wife. At issue during the conflict were the burgeoning piles of papers that were stacking up in my study, along the hall leading to the study, into the living room, up the stairs and even into the bedroom. You get the picture.

Although the prodigious paper collection was certainly capable of fending off most thrown objects, and possibly even a modest nuclear blast, I instead took refuge behind The Naked Scientists, and blamed that instead.

This is a live science radio show that I and a few colleagues set up almost by accident about ten years ago. The early days were a bit like good coffee – fun to enjoy with friends and nothing too serious. We would turn up at the radio studio on a Sunday evening and then spend a delicious hour talking about the breakthroughs that had made headlines that week, as well as playing some of our favourite songs.

Surprisingly, no one ever commented, at least in a bad way, about this rather erratic mix of science and chart music hosted by a bunch of junior doctors and scientists. And needless to say, The Chemical Brothers and Ph.D featured in the playlist on many occasions.

We continued in this vein for a number of years, moving to the BBC from the commercial circuit in the process. Then, in 2004, things really took off. The Winston Churchill Memorial Trust sponsored me to spend six months working in Australia with ABC Radio National. The idea was that I would be immersed in the world of high-stakes broadcasting to learn by osmosis how to make science programmes properly!

It was while I was in the midst of this 'sync-or-swim' experience that I wandered down the corridor from the office I worked in to where the Radio National breakfast team are based. I got chatting to them and they suggested maybe I'd like to provide some science coverage for them, which is something that I still do every Friday morning.

This was the catalyst that took me down the road to science reporting, and is ultimately how my house ended up resembling a paper-recycling warehouse. Since then, and pretty much every week, I read my way through a large proportion of the top papers published all over the world to distil the best science to discuss on the radio.

In the process I've found myself uncovering the most incredible science stories, charting new discoveries from outer space, new forms of life nourished by nothing but radioactivity and new ways to combat cancer and battle bacteria. Then, of course, there are the advances in technology that you wonder how we'd ever manage without, like the WiFi beer mat that knows when a glass is empty and automatically alerts the bar for a refill, and the radio-rucksack for pigeons so they can be used to get a bird's eye view of the smog levels hovering over cities.

So, although it was probably invalidating my household insurance on the grounds of fire risk, I kept all the material I read because I thought I might turn it into another book.

The argument with my wife that finally got me going began while we watched our baby daughter surfing down the stairs on a sea of PDFs. 'Okay, I'll write this stuff up,' was all I could say.

And this is it, another snapshot from the world of science. I am pleased to report that I now have a much cleaner and tidier house . . . and a paper-recycling tray.

FOOD

FOOD FOR THOUGHT AS RESEARCHERS UNCOVER APPETITE ACTIVATOR

SCIENTISTS LOOKING FOR ways to combat obesity have found an enzyme that activates an appetite-stimulating hormone. Researchers have been hungry to track down this target since blocking its action could help to stifle food cravings in people who are trying to lose weight.

Now, by using a genetic screening technique, University of Texas Southwestern researcher Mike Brown* has found it. Dubbed 'GOAT' (short for ghrelin O-acyltransferase), the new enzyme adds a chain of eight carbon atoms to a hormone called ghrelin, which is produced by the stomach and stimulates appetite. When injected into animals and human subjects, ghrelin boosts the urge to eat, while obese animals that have been genetically modified to lack the hormone or its receptor instead lose weight. But the hormone only exerts these appetite-inducing effects after it has been chemically modified by the addition of the carbon tail.

Previous research in fruit flies had shown that enzymes that add carbon chains to proteins like

* *Cell*, vol. 132, 8 February 2008

ghrelin all share a similar fingerprint sequence. So the researchers set about screening the mouse genetic code to look for signs of the same sequences. Using this approach they tracked down 11 genes and with one of them they hit the jackpot. By adding the gene to cells in a dish they were able to trigger the cells to produce the active form of ghrelin, demonstrating that they had identified the right gene.

Promisingly, ghrelin seems to be the only hormone in the body, at least as far as scientists can tell, that carries this eight-carbon unit activating tail, so scientists think that it should be relatively simple to block it and so combat the urge to binge. 'The discovery of GOAT opens the way to a search for chemical inhibitors that may be useful in controlling appetite,' say the researchers.

The proof as to whether it works or not will, of course, be in the pudding, or rather the lack of it!

ENERGY

KNEE-D MORE POWER?

SCIENTISTS IN CANADA and the US have developed a leg-driven knee-mounted generator that can power up to ten mobile phones just by harvesting the energy normally wasted when we walk. Simon Fraser University researcher Max Donelan and his colleagues* have based their invention on the 'generative-braking' principle used in some environmentally friendly cars. In this system instead of using the brakes to slow the car, the vehicle's momentum is instead turned into electricity that can then be used to start the vehicle moving again later.

The leg-generator works the same way. It consists of a modified knee brace which drives a small generator through a one-way clutch. It can be set to kick in only towards the end of a leg swing when muscles would need to be applied to stop the leg's movement. In this way the energy that it saves the wearer by slowing the leg is instead turned into electricity. In trials, with a device strapped to each leg, volunteers were easily able to produce electricity at the rate of 5 watts, which is six times greater than other footwear-based systems in

* *Science*, vol. 319. no. 5864, 8 February 2008

which the footfalls are used to generate power. The system also seems to compete very favourably with the next best alternative, which is a backpack in which a suspended load moves up and down to drive a generator as a person walks.

The team point out that most manual electricity generating devices occupy the full attention of the user but this system leaves the user free to attend to other tasks. The power it produces, say the researchers, could be used to run communication devices and personal computers, especially in countries with no reliable power supply.

HAIR CUTS CANCER MISDIAGNOSIS

SCIENTISTS IN AUSTRALIA have confirmed that samples of hair can be used to detect patients with breast cancer. Sydney-based researchers Gary Corino and Peter French* from Fermiscan Ltd repeated the work of another researcher, Veronica James, from the Australian National University**. She made the initial discovery in 1999 that when samples of hair are placed into a highly focused beam of X-rays, if the hair comes from a patient with breast cancer then the X-rays scatter from the hair with a different pattern than they do when hair from a healthy person is examined.

However, when other laboratories tried to repeat this work they were unable to reproduce the results, calling them into question. But now Corino and French have successfully done that, and shown that the technique does work so long as the hairs are carefully positioned in front of the X-ray beam, not under too much tension and the youngest (freshest) part of the hair is examined. Prepared like this, hair from healthy people produces an X-ray diffraction pattern, as it is

---

* *Nature*, vol. 398, 1999

** *International Journal of Cancer*, vol. 122, 2008

known, consisting of a series of arcs certain distances apart. Hair from breast cancer patients, on the other hand, has a circular ring superimposed over the arcs.

The test tends to generate false-positive results and 20% of healthy cases are incorrectly diagnosed as cancerous, but this doesn't rule out using it. Instead it means it could be a useful addition to the current system of breast assessment, since, in combination with other tests, it could help to improve the overall accuracy of screening.

As to why breast cancer alters the hair in this way, no one knows. The disease must be causing something to be added to the chemical structure of the hair in order to produce the ring that the researchers are seeing in their X-ray images. For now, hair-ever, its identity remains a mystery.

DISEASES AND DISORDERS

WINDOW INTO THE ALZHEIMER'S BRAIN

SCIENTISTS HAVE DEVELOPED a way to watch the pathological process that leads to Alzheimer's disease as it occurs in the brain. Harvard researcher Bradley Hyman and his colleagues* used a new imaging technique, called multiphoton microscopy, to peer into the brains of experimental mice that had been genetically preprogrammed to develop the rodent-equivalent of the disease. This new approach meant that the team could look different distances into the same brain regions of single animals over a number of weeks to look out for the appearance of any abnormalities.

Over the course of the study the researchers painstakingly matched up the locations of individual cells and blood vessels in each of the animal's brains as they looked for the development of amyloid plaques, the pathological hallmarks of Alzheimer's disease. These plaques are caused by a build-up of a protein called beta-amyloid, which is naturally produced in the brain but is normally broken down and removed. Why it accumulates, how quickly and how this affects the surrounding brain tissue was not known, but

* *Nature*, vol. 451, 7 February 2008

scientists had concluded that the process probably occurs slowly, as evidenced by the pace of progression of the condition in humans.

But the new study has changed all that, because the researchers were surprised to see the plaques forming and evolving very rapidly. Amyloid deposits could appear in less than 24 hours in a brain region previously free from lesions. Interestingly, once the plaques reached full size they stopped growing, possibly because other brain cells, called microglia, began to cluster around them. After this, adjacent nerve cells began to show structural abnormalities, answering another existing question, which was 'Which came first, the abnormal nerve cells or the amyloid plaques?'

Although it's early days, the new technique should provide researchers with a powerful new tool with which to probe the development of a disease which will inevitably affect one person in five in the developed world.

DISEASES AND DISORDERS

HUMANS 'PRIME-ATE' SUSPECTS FOR GIVING CHIMPS KILLER DISEASES

RECENT RESEARCH CARRIED out in West Africa has shown that humans are fatally infecting chimpanzees with our common cold viruses. Primatologist Sophie Kondgen, from the Robert Koch Institute in Berlin*, examined clinical samples collected from ill chimpanzees during respiratory disease outbreaks at a research site in the Cote d'Ivoire between 1999 and 2006. During the outbreaks, 92% of the animals at the study site developed symptoms and up to 20% of them subsequently died, with the majority of fatalities affecting very young animals.

The results of the analysis were surprising because the same two viral offenders were identified in all of the samples collected from the stricken animals. The culprits were two human viral infections called RSV – respiratory syncitial virus, and HMPV – human metapneumovirus, both of which cause coughs and wheezy symptoms in young children and more mild infections in adults. Intriguingly, tests on the viruses showed a series of characteristic genetic changes. The same molecular fingerprint had emerged

* *Current Biology*, vol. 18, Issue 4, 26 February 2008

shortly before in humans in South America, proving that the animals were almost certainly picking up the infections through human contact.

There are several possibilities to account for how the agents are spreading to the chimps, including poaching and tourism, but researchers themselves, because they spend long periods of time making observations close to the animals, are also prime suspects. Kondgen and her colleagues think that the viruses themselves are not necessarily fatal to the animals but instead make them susceptible to bacterial infections, triggering a fatal pneumonia.

But is the solution to ban humans from getting close to the animals? As the team point out, human presence is, in a way, good for the chimps' survival because it deters poachers, and the revenue from tourism encourages conservation. To tackle the problem, therefore, the researchers suggest mandatory vaccination programs for visitors, removal of human waste that could be infectious and the wearing of face masks capable of blocking the spread of viral particles.

HUMAN BIOLOGY

CLOCK WATCHING GETS UNDER THE SKIN

IN A BREAKTHROUGH that could help us to find better ways to combat sleep disorders and jetlag, scientists have discovered that skin and other cells can be used to tell the time of the body clock. Previously it was thought that the body uses just one main clock to keep track of time. This had been narrowed down to a cluster of nerve cells known as the suprachiasmatic nucleus, which is situated deep within the brain's hypothalamus. Cells in this region receive inputs from the eyes so that their time can be adjusted as the length of the day changes.

The clockwork mechanism itself consists of a group of (so far) eight genes, which turn each other on and off in sequence to produce a genetic domino effect that takes about 24 hours to complete. As the clock 'ticks', the behaviour of the nerve cells changes, leading to the secretion of hormones and other triggers, which directly affect the behaviour of different tissues around the body.

Zurich University researcher Steven Brown and his colleagues* have now found that the same patterns of

* *PNAS*, vol. 105, no. 5, 5 February 2008

activity also occur in skin cells, which use signals in the bloodstream from the master clock to set their own internal genetic timepieces. The team collected small skin samples containing cells known as fibroblasts from 28 healthy volunteers, 17 of whom described themselves as 'night owls' and the remaining 11 as 'larks'. The team used a virus to add a gene to the cells so that they would glow brightly whenever one of their clock genes turned on. The researchers then grew the cells in a dish and found the glowing marker gene periodically switching on and off with every 24-hour period. But most surprisingly, the time it took to do so was different depending upon whether the skin cells came from a 'lark' or an 'owl' human, with the larks having a significantly shorter period than their more nocturnal counterparts.

While interesting in itself, this finding shows that while the eyes might be the window to the soul, skin cells are a window to the mechanism of the body clock. 'This will certainly provide us with new insights into the workings of the body clock,' says 'clock-doc' Russell Foster, who works on circadian rhythms at Oxford University. 'There are large numbers of disorders that are also associated with sleep disturbance, like schizophrenia, depression, bipolar disorder and Alzheimer's. But how much clock disturbance contributes to the symptoms of these diseases we don't know. Techniques like this will show us and also enable scientists to develop better therapies.'

HUMAN BIOLOGY

'BRAINAL' BASIS OF ITCH-SCRATCHING SEEN FOR THE FIRST TIME

IN A BID to understand what triggers itching and why a scratch feels so good, US researchers have brain-scanned scratching volunteers to monitor the effects on their nervous systems. Gil Yosipovitch and his colleagues at Wake Forest University Medical School in North Carolina* recruited 13 volunteers who were exposed to 30-second bouts of scratching with a brush applied to their right calves over a 5-minute period. After each episode of scratching the volunteers were given a 30-second rest period. The researchers were then able to compare how brain activity changed when the subjects were actively scratched.

When they analysed the results the team were surprised to find that parts of the brain known to be concerned with processing unpleasant or aversive emotions, including a region called the anterior cingulate cortex (ACC), became much less active. Yosipovitch thinks that this is part of the reason why scratching feels so pleasurable, because it reduces the anxiety associated with itching. 'It's possible that scratching suppresses the emotional components of

* *Journal of Investigative Dermatology*, vol. 128, 2008

itch to bring about its relief,' he says.

The team also found that some parts of the brain increased their activity whenever scratching was applied, including the prefrontal cortex, which is concerned with anticipating rewards or benefits of an activity, and the insula cortex, which processes how the effects of an action will affect the body. This anticipated reward response might explain why patients with certain skin conditions, such as eczema, can scratch themselves to the point of drawing blood. 'It might explain the compulsion to continue scratching,' says Yosipovitch.

However, one drawback of the present study was that the participants were not themselves itchy at the time of the study, so the researchers are now returning to the scratching board to set up further experiments to test patients with chronic itching conditions to see if the results differ.

FOOD

SHED A TEAR FOR THE HUMBLE ONION

SCIENTISTS IN NEW ZEALAND have used GM (genetic modification) technology to develop onions that don't make you cry. Working with colleagues in Japan, Colin Eady, who is based at New Zealand Crop & Food Research*, has successfully silenced the gene responsible for producing the lacrimatory (tear-jerking) factor, which is a volatile chemical called syn-propanethial-S-oxide and is emitted by freshly cut onions.

Originally scientists thought that the chemical was pre-made within the vegetable, so preventing its production would probably interfere with the flavour of the onion. But in 2002 scientists in Japan successfully unpicked the chemical pathway responsible for its formation. Cutting into an onion releases the enzyme allinase, which breaks down a family of pungent chemicals in the onion called amino acid sulphoxides. Two further chemical steps, one of them achieved with the help of a second enzyme discovered by the Japanese and known as 'lacrimatory factor synthase', yields the syn-propanethial-S-oxide. It's this final step in the pathway that has been targeted by the

* *Plant Physiology*, vol. 147, no. 4, August 2008

researchers who have used a technique called 'gene silencing' to deactivate the enzyme.

The result is an onion that tastes great but doesn't reduce you to tears. In fact, according to Eady, preventing the sulphur compounds from being converted into the tearing agent redirects them into compounds responsible for flavour and health, so the process could even improve the taste of the onion. 'We anticipate that the health and flavour profiles will actually be enhanced by what we've done,' he said.

But cooks hungry to embrace the new eye-friendly foodstuff will have to wait for at least ten years before it is commercially available, says Eady. So in the meantime why not resort to folklore, which says that burning a candle close to where you are chopping an onion can help to burn off the irritant vapours. Or trying chopping your onions underwater – so long as you can hold your breath for long enough!

HUMAN BIOLOGY

NO PAIN, NO GAIN

SCIENTISTS ARE INVESTIGATING the possibility of using viruses to carry out painkilling gene therapy in the spinal cord. So far the technique, which is the brain-child of Andreas Beutler and his colleagues at Mount Sinai School of Medicine in New York*, has only been tested on rats, but the results look promising. The team have used an adeno-associated virus which has been genetically modified to replace the normal viral genetic material with a gene for one of the body's own natural painkilling chemicals called beta-endorphin, which works in the same way as morphine.

Ten billion particles of the modified virus were injected by lumbar puncture into the spinal cords of each of a group of rats. These animals had previously sustained nerve injuries that simulate the chronic pain experienced during certain human diseases. At the same time a second group of animals were injected with a control virus containing a harmless marker gene instead of the beta-endorphin gene. The animals were then followed up over three months to look for effects on their pain. Initially nothing happened,

* *Current Opinion in Molecular Therapeutics*, vol. 7, 2005

but then after one month the animals that had been injected with the beta-endorphin-containing virus began to show significant improvements in their pain, which continued until the end of the experiment two months later. The control animals, on the other hand, remained unchanged.

The team think that their approach could make life much more comfortable for human patients experiencing severe pain, such as those with terminal cancers. This is because effective pain control is a delicate balancing act: doctors need to walk a tightrope between patient comfort and side effects. 'Chronic pain patients often do not experience satisfactory pain relief from available treatments due to poor efficacy or intolerable side effects like extreme sleepiness, mental clouding and hallucinations,' says Beutler. 'But targeted gene therapy will likely avoid the unwanted side effects associated with opioid painkillers such as morphine.'

HUMAN BIOLOGY

ORGAN TRANSPLANTS WITHOUT REJECTION

FOUR OUT OF five patients participating in a recent kidney transplant trial in America have been able to stop using immune-suppressing drugs. The patients, aged 22–46 years old, had all developed kidney failure for different reasons. Prior to receiving the donor organ each of the patients received a course of drugs to partially destroy their bone marrow and the T lymphocytes, white blood cells that recognise and reject foreign tissue. After this 'conditioning' regimen was complete the patients underwent a combined transplant during which they received the new donor kidney and were also injected with their donor's bone marrow cells.

The results have been striking. Between 9 and 14 months after each transplant all but one of the patients have been able to stop taking any immunosuppressive drugs, and their kidneys remain healthy without any signs of rejection up to 5 years later. The scientists think that the transplantation of the donor bone marrow triggers the recipient's body to develop 'tolerance' to the foreign organ, possibly by re-educating the recipient's white blood cells into ignoring what was previously viewed as hostile.

According to senior study author David Sachs from the Massachusetts General Hospital Transplantation Biology Research Center in Boston*, 'We are very encouraged by our initial success in inducing tolerance. While we need to study this approach in a larger group of patients before it is ready for broad clinical use, this is the first time that tolerance to a series of mismatched transplants has been intentionally and successfully induced.'

The findings will almost certainly come as welcome news to the thousands of patients who face a twice-weekly trek to their nearest hospital for dialysis while they wait for a suitable donor organ to be found.

---

* *New England Journal of Medicine*, vol. 358, 2008

HUMAN BIOLOGY

PREGNANCY–CAFFEINE COMBO BAD IDEA

RESEARCHERS HAVE SHOWN that caffeine exposure is linked to an increase risk of miscarriage. De-Kun Li, from Californian health insurer Kaiser-Permanente*, followed 1063 women during the first 20 weeks of their pregnancies or until they had a miscarriage. Amongst the details collected, the participants were asked about their caffeine consumption.

The data is published in the *American Journal of Obstetrics and Gynecology* and shows that women drinking more than 200 milligrams of caffeine per day – the equivalent to just over one cup of coffee – had a two-fold rise in their miscarriage risk. But it wasn't just coffee that was to blame – all caffeinated beverages including fizzy drinks, tea and even hot chocolate had the same effect.

According to Li, caffeine might be triggering vaso-constriction – blood vessel narrowing – in the placenta, cutting blood flow and the delivery of oxygen and collection of waste products from the developing foetus. Alternatively, he says, caffeine may be directly

---

* *American Journal of Obstetrics and Gynecology*, vol. 198, Issue 3, 2008

toxic to a foetus. Either way, he suggests that women avoid caffeine altogether during pregnancy.

GEOLOGY

SCIENTISTS MAKE WAVES IN SEISMIC VOLCANO FORECAST

THEY SAY THAT you can hear the sound of the sea in a shell held close to your ear. Sadly that's a myth, but the sound of the sea can certainly be heard by listening to a volcano, where it can be used to predict an eruption, say French researchers.

Writing in the journal *Nature Geoscience*, Grenoble University's Florent Brenguier and colleagues* have found that when waves hit a beach they produce low frequency seismic shock waves which travel long distances through the Earth's crust. But if they pass through the magma chamber of a volcano that is about to become active the waves change in a highly characteristic way, warning researchers that an eruption is imminent.

The team used the technique to correctly predict, up to ten days in advance, the 2006 and 2007 eruptions of the Piton de la Fornaise volcano on the island of Réunion in the Indian Ocean. According to Brenguier, an array of 20 sensors around the volcano is sufficient to give a detailed picture of what is happening below ground at any depth. But is the research

---

* *Nature Geoscience*, vol.1, January 2008

tied to the ocean? No, says Brenguier; other sources of vibration including natural seismic sources like the wind pummelling the ground are equally effective. Traffic noise could also be used, but its higher frequency means that it tends to be less penetrating than waves from the sea.

NON-HUMAN BIOLOGY

RED ANT, DEAD ANT . . . PARASITE MAKES ANTS RESEMBLE FRUIT

SCIENTISTS HAVE DISCOVERED a parasite that causes ants to resemble fruits, which turns them into choice morsels for birds. Working on tree-dwelling ants in the Panama and Peruvian Amazon, University of Arkansas researcher Steve Yanoviak and his colleagues* initially thought that the ants scuttling past with large red shiny abdomens were part of a new species. But careful examination of the red-reared insects, which are actually a well-known species called *Cephalotes atratus*, revealed that the ants were crammed with hundreds of parasite eggs, each of which contained a tiny nematode worm. The yellow colour of the parasites, together with the natural dark colour of the ant's body, is what makes the insects turn red.

The team also found that infected individuals tend to carry their infected abdomens, known as gasters, pointing upwards, which makes them much more prominent, and that the abdomen is much easier to detach from the body, possibly because the infection also thins the tissue making up the ant's outer skeleton. The ants also seem to become less aggressive under

---

* *The American Naturalist*, vol. 171, 2008

the influence of the parasite, and cease to produce pheromonal warnings whenever predators approach.

Together with their colourful complexion, these changes make them much more conspicuous to birds, which mistake them for the small fruits prevalent in the tree canopy. The parasite eggs then pass unharmed through the bird's digestive tract and the infectious cycle is completed when uninfected ants subsequently scavenge the bird's faeces to feed to their young, which then become infected. 'It's phenomenal that these nematodes actually turn the ants bright red, and that they look so much like the fruits in the forest canopy,' says Yanoviak. 'When you see them in the sunlight, it's remarkable.' The findings have been published in the journal *The American Naturalist*.

CHEMISTRY

MEGA-CELL: RESEARCHERS BUILD BETTER BATTERIES

US SCIENTISTS HAVE come up with a way to make lithium batteries last ten times longer, which means a laptop could last all day on just one charge. Stanford's Yi Cui and his colleagues* made the breakthrough by using a forest of tiny silicon nanowires, which act as the positive electrode, known as the anode, inside the battery.

The present generation of batteries use graphite electrodes. These store electrical energy by soaking up and later releasing lithium as the battery charges and discharges; but the amount of lithium they can absorb is limited and this reduces the overall capacity of the battery. Silicon has a much higher charge capacity, so using it should enable scientists to build better batteries – but there's a catch. When silicon soaks up lithium inside the battery, it swells up by 400% and then shrinks again when the lithium is later released. This cycle occurs every time the battery is charged and discharged, and very quickly the silicon becomes 'pulverised' into tiny fragments which lose contact with their parent electrode and cause the battery capacity to

* *Nature Nanotechnology*, vol. 3, 2008

fall. Instead Cui and his team used a technique called VLS, short for vapour–liquid–solid, to grow clusters of short silicon wires each about one ten-thousandth of a millimetre in diameter. When the new electrodes were tested as part of a lithium ion cell, the increased surface area helped to boost the amount of charge that the battery could hold, the cells worked with over 90% efficiency, and the electrodes were also capable of handling currents five times greater than those tolerated by existing graphite electrodes. Moreover, when the team used X-rays to study the crystal structure of the silicon nanowires when they were charged and discharged, they found that the wires swelled and shrank without signs of pulverisation.

'We are working on scaling up and evaluating the cost of our technology,' Cui said. Cui has filed a patent on the technology, which is published in the journal *Nature Nanotechnology*. He expects the battery to be commercialised and available within 'several years'.

NON-HUMAN BIOLOGY

SQUIRRELS FAKING IT WITH THEIR NUTS

SQUIRRELS SEQUESTERING NUTS and other nutritious morsels is a common sight in autumn. But, according to US researchers, all is not what it seems, because up to 20% of the time squirrels are faking it and their elaborate burial displays are just to put other hungry individuals off the scent.

Writing in *Animal Behaviour*, Wilkes University researcher Michael Steele and his colleagues* watched the antics of over 100 grey squirrels and compared how the animals' behaviour changed when the team began pilfering their cached food. The squirrels reacted to the food thefts by doubling the number of fake burial displays they performed, hiding food in locations that were much harder to raid, and also eating more of it straight away.

These results seem to suggest that squirrels might behave like western scrub jays, which also hide food and pay close attention to who is watching them as they do so. If they find themselves under scrutiny, they return later and move their food caches. This has led some researchers to propose that the birds have a

* *Animal Behaviour*, vol. 75, Issue 2, February 2008

'theory of mind' – in other words, an understanding of the intention to steal – so now it looks like squirrels fall into the same mould!

HUMAN BIOLOGY

DO FOUR THINGS, GAIN 14 YEARS

RESEARCHERS AT CAMBRIDGE University have found the secret to living 14 years longer – don't smoke, drink moderate amounts of alcohol, take small amounts of exercise and, above all, eat your greens!

Professor Kay-Tee Khaw and her colleagues* followed up 20,000 men and women aged between 45 and 71 years old who were recruited between 1994 and 1997. The study subjects filled in a simple questionnaire about their lifestyles and earned a point for every positive answer to being a non-smoker, light drinker, and regular exerciser. They also earned a point if a blood test revealed a vitamin C level consistent with eating about five daily portions of fruit and vegetables. Deaths amongst the volunteers were then recorded until 2006.

When they analysed the results the researchers found that, over an average period of eleven years, people who scored zero were four times more likely to die than participants who had scored four points. They also found that people scoring zero had about the same risk of dying as someone 14 years older than them who had a score of 4 points.

---

* *PLoS Medicine*, vol. 5, no. 1, January 2008

So the moral of this story is quit smoking, drink in moderation, take the stairs not the lift and follow the five-a-day plan (that's portions of fruit and veg, not hamburgers), and you might live 14 years longer.

NON-HUMAN BIOLOGY

BIOLOGICAL DOMINO EFFECT: NO ELEPHANTS = NO ANTS = NO TREES

US ECOLOGISTS HAVE highlighted the delicate balance at work in nature with a study showing that the disappearance of elephants has knock-on effects on ants and ultimately the survival of trees.

Working in Kenya, University of Florida researcher Todd Palmer and his colleagues* simulated the extinction of large grazing mammals such as elephants and giraffes by using fences to exclude the animals from patches of ground. Then, for the next 10 years they monitored the effects on a species of acacia tree, known as whistling thorns (*Acacia drepanolobium*) and the ants that inhabit the trees. The trees normally produce specialised hollow thorns and secretions of nectar to attract certain ant species which, in return for their thorny lodgings and the food, vigorously defend their host tree against other parasites and browsing animals.

But once the large animals were excluded, things began to change. The species of ants inhabiting the trees changed, and the ant colony sizes reduced dramatically. The trees also became less healthy, grew slower

---

* *Science*, vol. 319, no. 5860, 11 January 2008

and were more than twice as likely to die compared with normally grazed trees. So what provoked this dramatic change? The lack of grazing by elephants and other large animals resulted in the trees reducing their production of ant-incentives, including the hollow thorns for them to live in and nectar rewards. Consequently the ants that would normally defend the trees, a species called *Crematogaster mimosae*, were lost and replaced with other ant species including one called *Crematogaster sjostedti*, which lives in cavities created in tree stems by long-horned beetles, which the ants themselves encourage to colonise the tree. But while it provides a home for the ants, the wood-weakening effect of the beetle larvae kills the trees.

So, unpredictable as it would seem, if elephants and other large African mammals were to go extinct, trees and even ants would immediately suffer, highlighting the fragility and interdependence of every species in nature.

HUMAN BIOLOGY

WHOLEHEARTED ATTEMPT TO REPAIR BROKEN HEARTS

SCIENTISTS HAVE SUCCESSFULLY grown a new heart in the laboratory.

The researchers, from Harvard and the University of Minnesota, first used a process called organ decellularisation to remove all of the existing cells from the hearts of dead rats; this was achieved by perfusing the organ with a mixture of detergents which broke open and washed away the dead heart cells but left behind a scaffolding of connective tissue from which healthy new cells could be hung. The scaffolding was then bathed in a nutrient culture solution and injected with stem cells harvested from the hearts of newborn rats.

Over days to weeks the stem cells divided and began to colonise the heart scaffolding, turning into new heart cells as they did so. Within eight days after the cells were injected, the team could see pulsations as the developing heart cells began to beat, and the cells also responded to the administration of adrenaline-like drugs that affect the behaviour of normal heart cells.

The researchers, led by Doris Taylor*, now hope that the technique could be used to repair damaged

* *Nature Medicine*, vol. 14, 2008

hearts in humans. 'We used immature heart cells in this version as a proof of concept,' says Taylor. 'Going forward, our goal is to use a patient's stem cells to build a new heart.' This would help to avoid the problems of immune rejection of donated organs.

But can the technique be scaled up from a rat to something the size of a human heart? Yes, say the researchers who have now successfully used the technique on a pig's heart, which is similar in size to our own. According to team member Harald Ott, 'we just took nature's own building blocks to build a new organ.'

HUMAN BIOLOGY

RESEARCHERS UNCOVER SMOKING GUN IN TEENAGERS' BRAINS

RESEARCHERS AT YALE Medical School in the US have discovered that smoking during adolescence affects the development of the brain. In particular, the regions of the nervous system that control how we pay attention to sounds and visual stimuli are altered, making the smoker more prone to being distracted.

Leslie Jacobsen and her colleagues* brain-scanned almost 67 teenage volunteers, half of whom smoked, half of whom didn't. Roughly half of the volunteers also had mothers who smoked during their pregnancy. The scans revealed that both pre-natal and teenage exposure to tobacco smoke were associated with changes in the thalamocortical and corticofugal nerve pathways, which run in a region of the brain called the internal capsule and are concerned with how we pay attention to sounds and other stimuli. The findings agree with a study carried out previously by Jacobsen and her team in which they showed a link between smoking and a reduced ability to focus on a piece auditory or visual information without being distracted by other things going on at the same time.

---

* *The Journal of Neuroscience*, vol. 27, no. 49, 2007

The researchers suspect that nicotine is probably to blame for the effect because experiments carried out previously on animals have shown that it can affect the development of nerve connections that control how information flows into the brain.

The next step, says Jacobsen, is to brain-scan teenagers who have quit smoking. This will reveal whether the differences she has picked up are reversible or not.

METEOROLOGY

SCIENTISTS DUST OFF HURRICANE WARNING THEORY

SCIENTISTS AT THE NASA Goddard Space Flight Center in Maryland, US, may have a new tool to offer weathermen – a way to predict a bad hurricane season.

Hurricanes originate in the Caribbean and western Atlantic when high sea temperatures warm the overlying air sufficiently and trigger strong winds as the hot air starts to rise. But researchers William Lau and Kyu-Myong Kim wondered whether the 200 million tonnes of sand and dust blown every year into the atmosphere from the Sahara desert could have a shielding effect, cutting down the amount of sunlight hitting the sea and therefore reducing ocean temperatures and the numbers of hurricanes. To find out, they compared satellite images of dust clouds from the Sahara with sea temperatures in the summers of 2005 and 2006. In 2006 the sea temperatures were lower, no hurricanes hit the land that year, and sure enough there was a larger than average dust cloud. The effect is most marked in summer, between June and August. Before then the dust is too scarce in the atmosphere to make much difference.

The research, which is published in the journal

*Geophysical Research Letters**, could help forecasters to predict future storms with greater accuracy. Indeed, a Colorado State University team that produces quarterly hurricane predictions plans to include Saharan dust figures in their weather models in future. According to the group leader Phil Klotzbach, 'The dust information will be quite useful for our final seasonal forecast in August.'

* *Geophysical Research Letters*, vol. 33, 2006

ECOLOGY

POLLUTION BRAIN DRAIN

SCIENTISTS IN THE US have found evidence to suggest that exposure to sooty traffic fumes is costing children up to 3 IQ points of intelligence. Writing in the *American Journal of Epidemiology** Harvard researcher Shakira Franco Suglia studied 200 children living in Boston, US, and found that those with the highest exposure to traffic pollution performed less well on intelligence tests than their cleaner breathing counterparts, even after taking factors such as social class into account.

This puts traffic pollution on par with lead in terms of its potential to stunt a child's brain development, although the researchers aren't yet sure exactly which components of exhaust fumes are responsible. But previous studies on animals suggest that particulates (tiny invisible particles smaller than the body's own cells) are likely to be the culprits. These can enter the bloodstream through the lungs, but there is also evidence that they can penetrate into the central nervous system via the nose, travelling along the olfactory nerves that carry the sense of smell into the brain.

* *American Journal of Epidemiology*, 1 February 2008

Once there they can directly damage cells and also trigger inflammation that further injures the brain.

This not as unlikely as it sounds. Studies on dogs living on the smoggy streets of Mexico City have been shown that animals exposed to the most pollution develop brain damage similar to the changes seen in humans with Alzheimer's disease.

Now researchers need to track down what components of traffic pollution are responsible and determine how to remove them from the exhaust pipes of the millions of cars and trucks pounding the tarmac of big cities in every part of the world . . . so don't hold your breath. Or on second thoughts, do hold your breath!

PREVENT DIABETES WITH A GOOD NIGHT'S SLEEP

SCIENTISTS HAVE DISCOVERED that a component of your nocturnal nap, known as slow wave sleep, is critical to helping the body to regulate sugar levels and stave off diabetes.

Writing in the journal *PNAS*, Esra Tasali and her colleagues* report that, at the University of Chicago Medical Center, they recruited nine healthy young volunteers and monitored their sleep. The subjects were rigged up to a monitor which played noise from a speaker whenever their brain waves showed that they had entered slow wave sleep, which is the deepest phase of sleep. The sounds were just loud enough to disrupt the subjects' sleep, but without waking them up. This occurred about 250 times during the night for each subject and enabled the researchers to cut the amount of slow wave sleep amongst the volunteers by 90%.

The effect is similar to ageing the brain by 40 years, says Tasali. 'In this experiment we gave people in their 20s the sleep of someone in their 60s.' When the volunteers woke up each morning the researchers administered a small amount of glucose intravenously

* *PNAS*, vol. 105, no. 3, 22 January 2008

and then monitored blood sugar and insulin levels. The researchers were surprised to see that by the end of the study the volunteers had become 25% less sensitive to insulin and their glucose levels were 23% higher.

The team point out that their findings might help to explain why obese and some elderly individuals develop type II diabetes. Both old age and obesity are associated with poor sleep, which could be responsible, at least in part, for triggering some of the cases of diabetes seen in these individuals. 'Our findings raise the question of whether age-related changes in sleep quality contribute to the development of these metabolic alterations,' the team says.

HUMAN BIOLOGY

GETTING STEM CELLS TO REPAIR A BROKEN HEART

SCIENTISTS HAVE DISCOVERED how stem cells can be used to prevent a fatal complication of heart attacks known as ventricular tachycardias. These disturbances of heart rhythm are the most common cause of death following a heart attack and they're caused by abnormal electrical activity in the part of the heart that's been damaged. But now, writing in *Nature*, Bonn University researcher Wilhelm Roell and his colleagues* say that they think they know why and how to use stem cells to prevent the problem from occurring. To simulate the damage done by a heart attack the team used a cold probe to make small lesions in the hearts of experimental mice. The animals were then divided into two groups. In one group the damaged region of the heart was immediately injected with stem cells that had been collected from either the developing leg muscles or hearts of embryonic donor mice. To make them easy to identify the donor cells were labelled with a green dye so that the team could follow where they went in the recipient animals. A second group of animals were not injected with stem cells and were followed up as controls.

---

* *Nature*, vol. 450, 6 December 2007

The researchers then tested the susceptibility of the animals to developing potentially fatal ventricular tachycardias (VT). The problem could be provoked in 100% of the control animals and the animals that received injections of muscle stem cells, but in only 39% of the animals that had received heart stem cells, making this group of animals no more susceptible to VT than healthy mice. When they examined the animals' hearts the team found that the heart-derived stem cells, but not the muscle stem cells, were producing high levels of a protein called connexin 43, which electrically links heart cells together. In this way the donor cells were wiring themselves into the damaged heart and producing an electrical bridge across the injured area, so that the heart rhythm could continue to be conducted normally. To prove that this was the case the researchers then used genetic techniques to make muscle stem cells that were also capable of producing connexin 43. This time, when the cells were injected, the mice were much less vulnerable to VTs.

On the basis of these results the team think that this technique could be used to save lives following heart attacks. The connexin gene could be added to stem cells collected from a patient who has just suffered a heart attack, and these cells could be injected into the injured heart, helping to ensure that it remains electrically stable.

NON-HUMAN BIOLOGY

AREA X-FACTOR KEY TO SINGING THE WRONG SONG

SCIENTISTS WORKING WITH songbirds have uncovered the genetic basis of human speech. Sebastian Haesler and his colleagues, writing in the journal *PLoS Biology**, have shown that a gene called FoxP2, which is expressed in a part of the songbird's brain known as area X, is essential for young birds to learn their songs. Levels of this gene increase during the time when the birds are acquiring and adapting their songs.

To find out whether the gene is actually responsible for the process the team made a genetically modified virus containing the genetic 'mirror image' of FoxP2. When this virus was injected into a young bird's brain it had the effect of switching off the FoxP2 gene and preventing it from being expressed in that part of the nervous system. And when the team injected young, 3-week-old zebra finches with the virus and then paired them up with an older 'tutor' bird they managed at best to learn only incomplete and inaccurate versions of their teachers' songs. Un-injected control birds, or birds injected with a virus that left the FoxP2 gene unaffected, imitated their tutors much more faithfully.

---

* *PLoS Biology*, vol. 5, no. 12

The results bear a striking similarity to a human condition that is caused by inheriting a defective version of the same gene that was knocked out in the birds – FoxP2. Subjects with the condition have difficulties with the production of fluent speech and grammar, suggesting that the present study might hold the key to better understanding the neural basis of how we learn and produce language.

COMPUTER TECHNOLOGY

COMPUTERS CRACK DOWN ON DRUG SIDE EFFECTS

AT ONE TIME the only way to find out whether a potential drug might be safe was to test it on animals and then on humans. And it was often only after it had been licensed and begun to fly off doctors' prescriptions pads that rare, and sometimes deadly, side effects were uncovered. Now scientists have developed a computer program that can screen potential drug molecules for possible side effects before they've even left the test tube, a move which could save pharmaceutical companies a fortune.

The work is the brainchild of UC San Diego computer scientist Philip Bourne*. He and his colleagues have based their approach on a technique already employed by drug companies to first identify chemicals that might have therapeutic effects. Pharmaceutical developers run simulations comparing structures of molecules that might be linked to certain diseases in the body with chemical compounds they have made. They're looking for signs that their compound might be able to affect their chosen molecular 'target'. So what the San Diego team have done is to develop a system

* *PLoS Computational Biology*, vol. 3, no. 11

that compares drug molecules against the structures of large numbers of human proteins – 800 in the present study – looking for interactions.

As a proof of concept they tested their approach on the breast cancer drug tamoxifen and identified a hit. The system predicted an effect on a protein that pumps calcium into and out of cells. And this fits perfectly because tamoxifen is known to cause heart problems in some patients, and heart cells rely on calcium to contract. But the approach, published in *PLoS Computational Biology*, still has some way to go. The human genome encodes at least 25 000 different proteins, many of which have not yet been fully or even partially characterised. However, it's early days and an encouraging start. Bourne points out, 'There is a lot of potential for this process.'

ECOLOGY

BIODEGRADABLE PLASTICS? IT'S IN THE BAG, SAY RESEARCHERS

SCIENTISTS MAY HAVE come up with a solution to the problem of how to tackle waste plastic pollution. They've borrowed from biology and then tweaked the recipe slightly to come up with a class of plastics that completely break down in the environment in a matter of weeks.

Cornell University's Emmauel Gianellis and his team* used a natural polymer called poly 3-hydroxy-butyrate (PHB), which is made by some microbes to store energy, in effect as the bacterial equivalent of body fat. But while PHB is degradable and biocompatible, its other chemical and mechanical properties are less ideal. Now the Cornell team have found that by adding some silica (clay) nanoparticles they can significantly improve the degradability, barrier and chemical properties of the material, including making it tolerate higher temperatures, which makes it suitable for a larger range of potential applications.

The new composite behaved as a clear plastic polymer but broke down almost completely in compost, at room temperature, within 7 weeks. And

* *Biomacromolecules,* vol. 8, no. 11, 2007

what makes it very attractive as a material is that it showed almost no degradation for the first 4 weeks, before then breaking down very rapidly. The team think that the more rapid breakdown is due to the spaces between the particles being more vulnerable to chemical attack and then consumption by microbes.

GENETICS

WILT-NOT WANT-NOT

Scientists in the US have created a strain of GM drought-resistant plants which, they say, could help to preserve crop yields and combat the effects of climate change. The experimental tobacco plants created by UC Davis researcher Eduardo Blumwald and his colleagues* grew normally under ideal conditions, could still grow well when given only a fraction of their normal requirements and were even able to tolerate 15 days without water altogether. Unmodified 'control' tobacco plants exposed to the same conditions, however, all died.

The team created the plants by blocking a process known as 'leaf senescence' in which plants artificially age and drop their leaves in order to reduce their leaf area and hence their water requirements. While this can help to preserve the growing cycle of the plant it reduces the yield of an annual crop. So to prevent this from occurring the researchers used genetic techniques to switch on a gene called IPT – isopentenyltransferase – whenever and wherever the plant tissues became stressed by a lack of water. IPT boosts the production of a growth-related

* *PNAS*, vol. 104, 2007

gene called cytokinin (CK), which helps to protect leaves against the effects of water shortage. Most encouragingly, the transgenic crops also continued to grow well under conditions of severe water restriction. Grown side by side with unmodified control plants and given only a third (0.3 litres) of their normal water requirement for 4 months to simulate a prolonged dry spell, the modified plants dropped their yield by only 8-14% compared with a 60% loss of yield for the unmodified plants.

The team hope that their results could be used to produce low-water tolerant crops that could be used to make arid land more productive, or enable farmers to save irrigation water while maintaining higher crop yields.

MEDICINE

DOCTORS REFLECT ON MIRRORS AS A THERAPY FOR PAIN

DOCTORS IN AMERICA have shown that a mirror is an effective therapy for the 'phantom' pain experienced by 90% of patients undergoing limb amputation. Writing in the *New England Journal of Medicine*, Bethesda-based researcher Jack Tsao and his colleagues* describe how they recruited 22 lower limb amputees. The volunteers were randomly divided into three groups; in one group, the 'mirror group', the reflection of the subject's intact limb was used to create the illusion that the missing body part was still present. The patients were told to try to make movements with their absent limb by watching the reflection of their intact leg making the same movement. In the second 'control' group the mirror was covered with an opaque sheet, while the third group of patients were instructed just to imagine making movements with their amputated leg.

The results were shocking: after four weeks of treatment 100% of the patients in the mirror group reported that their pain had improved. Patients in the other two groups fared less well. Among the covered-

---

* *New England Journal of Medicine*, vol. 357, no. 21, 22 November 2007

mirror volunteers, 50% felt that their pain had worsened (although one patient reported improvement), and the visualisation group fared worst of all: 67% said their pain had intensified. Next the researchers switched the patients from the visualisation and covered-mirror groups to mirror therapy, and their pain improved dramatically, becoming similar to that reported by the first group of mirror patients.

The researchers aren't certain how the technique works but suspect that a population of appropriately termed 'mirror' neurones in the brain may be responsible. These nerve cells fire when a person makes a movement or watches another individual performing the same movement, so it may be that the reflected image fools these cells into responding as though the amputated limb were still present, altering the pain perception.

PHYSICS

TERROR-HERTZ SCANNERS

SCIENTISTS HAVE DEVELOPED a compact way to produce a safe form of electromagnetic waves – known as 'T-rays' – that could revolutionise airport security scanners and also give rise to powerful new diagnostic techniques. The terahertz rays are radio waves with a frequency 1000 times higher than microwaves. They are attractive as a medical and security imaging system because the waves are harmless – they are not sufficiently powerful to ionise or knock electrons off other chemicals, which means that they cannot damage DNA or trigger cancers. But they are highly penetrating, which means that they can be used to 'see' through varying thicknesses of skin, leather, cardboard and clothing, although they are stopped by water and metals. They can also be used to 'chemically fingerprint' materials because certain substances absorb specific characteristic frequencies of the waves.

Unfortunately these waves were very difficult to produce because there were no materials available to generate the required frequencies in a portable fashion. But now a team of researchers in the US, Turkey and

Japan* have taken advantage of a crystal trick called a Josephson Junction to produce a highly compact terahertz radiation source. These microscopic junctions, which are 10,000 times thinner than a human hair, consist of a sandwich of superconducting copper oxide insulated by layers of bismuth oxide and strontium oxide. When a voltage is applied across the junction it spits out photons of light (electromagnetic waves) at terahertz frequencies, although the intensity is very low.

So, to produce a more powerful signal, the researchers stacked up 1000 of these junctions on top of each other. Then, to make all of the junctions emit photons at the same time (so that the signal would add together), they arranged the junctions into resonant cavities – the quantum effect of an echo chamber. In this way the signal builds up and sustains itself, while also emitting some electromagnetic radiation that can be used. And by varying the length and thickness of the cavities the team were able to generate frequencies from 0.4 to 0.8 terahertz at a power of 0.5 microwatts. To push up the output power yet further to an ideal 1 milliwatt, which would have useful applications, the team are now experimenting with arranging the cavities into arrays so that their outputs can be added together.

---

* *Science*, vol. 318, no. 5854, 23 November 2007

HUMAN BIOLOGY

MINTY FRESH MAGNOLIA

RESEARCHERS HAVE FOUND that the key to fresh breath might lie in an extract of magnolia bark. Michael Greenberg and his colleagues* found that, in the test tube, two key chemicals in the bark, magnolol and honokiol, could kill more than 99.9% of the bacteria that cause bad breath in just 5 minutes.

To find out whether the extracts could also work in the mouth, the researchers collected saliva samples from 9 volunteers 1 hour after lunch and then at 30 and 60 minutes after eating a mint laced with the chemicals or at 40 minutes after chewing a piece of chewing gum, containing the same chemicals. In separate trials the subjects also chewed control gum and mints that did not contain any magnolia extracts. The saliva was then incubated on bacterial growth media and the numbers of bacterial colonies that grew were counted.

The results were impressive. The mints reduced salivary bacteria by 62% at 30 minutes and 33.8% at 60 minutes, while the chewing gum cut bacteria by 43% at 40 minutes. By comparison, the placebo mint cut bacterial levels by 3.5% at 30 minutes and

---

* *Journal of Agriculture and Food Chemistry*, vol. 55, no. 23

increased bacteria by 50% at 60 minutes. The placebo chewing gum cut bacteria by 18%. The researchers point out that this could represent an effective way to cut oral bacteria without the risks of side effects and 'rebound halitosis' (subsequently worse oral malodour!) associated with other breath-freshening strategies. For instance, triclosan has been found to react with chlorine in drinking water to form toxic substances, chlorhexidine-based solutions have been found to stain teeth, while alcohol-based remedies can dry the mouth and make the problem worse.

The team behind the results are based at Wrigley's, so it shouldn't be too long before magnolia gum hits the shelves; but whether it will stick is another matter.

COMPUTER TECHNOLOGY

HONEYBEE DANCE HELPS TO KEEP WEBSERVERS BUZZING

SCIENTISTS FROM AMERICA and Oxford have taken a leaf out of the bee's book in order to boost the efficiency of servers that power websites. Craig Tovey and Sunil Nakrani* were discussing the strategy that bees use to find food when they realised that the clever communication system used by the insects could also help to build better computer systems. Hives have a limited number of bees and a limited amount of energy and time with which to gather nectar, so they have to be highly efficient in where they choose to deploy their efforts. Sources of nectar can also change rapidly, with new sources becoming available and other sources drying up.

To get around the problem the insects alert each other to potential sources of food with the use of a waggle dance performed in front of other bees inside the hive. The direction of the dance indicates in which direction the food lies, the number of turns in the dance conveys how far it is, and the overall length of the dance the sweetness of the nectar. The watching bees learn the dance and having checked out the nectar

---

* *Bioinspiration & Biomimetics*, vol. 2

source for themselves then return and teach other bees. But in the meantime, if another nectar source has turned out to be more fruitful, then the original dance is collectively forgotten and the workers pick up the new moves.

So how does this apply to the internet? Well, different functions on some websites are driven by different webservers, and if a particular area of a site suddenly experiences heavy demand (the equivalent of bees finding a rich source of nectar), a queue can build up as the server supplying that function struggles to cope with the load; other servers, meanwhile, might well be standing idle. But the bee-based system developed by Tovey and Nakrani uses a virtual dance floor where servers advertise how busy they are to each other. The length of the ad determines the magnitude of the load in a particular area. So when an available server 'sees' an ad it helps out, helping to balance out the load and apportion resources where they are most needed to keep websites buzzing along at top speed.

HUMAN BIOLOGY

RESEARCH NOT SO PASSIVE WHEN IT COMES TO SMOKE DAMAGE

SCIENTISTS IN THE US have produced strong evidence for the harmful effects of passive smoke inhalation. Chengbo Wang and his colleagues at the Children's Hospital of Philadelphia* made the discovery by using a new imaging technique called global helium-3 diffusion magnetic resonance imaging (MRI), to test the lungs of 60 volunteers aged between 41 and 79 years, including 15 current or former smokers and 45 lifelong non-smokers, half of whom had been exposed to high levels of passive smoking in the past (indicated by living with a smoker for more than 10 years).

The subjects breathed in a small amount of a form of helium called helium-3, and the MRI scanner was used to monitor how the atoms moved around inside the subjects' airways, a measure called the ADC or apparent diffusion coefficient. In healthy non-smokers the air spaces in the lungs are small and compact, which helps the blood to efficiently collect oxygen and shed carbon dioxide. This also means that the helium atoms don't tend to move very far. But smoking causes lung tissue to break down, which is

* www.rsna.org/

known as emphysematous change, and makes the air spaces larger. When these changes are present the helium atoms can move much greater distances, and this can be picked up by the scanner. Indeed the researchers found that almost one-third of the non-smokers with high exposure to second-hand smoke had structural changes in their lungs similar to the smokers, suggesting that damage was present.

'To our knowledge, this is the first imaging study to find lung damage in non-smokers heavily exposed to second-hand smoke,' said Wang. 'We hope our work strengthens the efforts of legislators and policymakers to limit public exposure to second-hand smoke.'

NON-HUMAN BIOLOGY

HOW MICE 'PEE-CEIVE' WHO THEY'RE RELATED TO

SCIENTISTS HAVE UNCOVERED how mice identify their relatives and so choose not to mate with them – by the smell of their urine! Previous studies in mice and in humans, such as the smelly T-shirt test, had suggested that animals try to choose mates with a selection of immune system 'MHC' genes that are as different as possible from their own. But now Liverpool University's Janet Hurst and her colleagues* have found that this is not true and instead mice rely on a diverse set of smelly chemicals in their urine called MUPs – major urinary proteins – which are encoded by the animal's DNA. These give a mouse a chemical barcode, which it can compare against a potential mate. If the smells match then the two animals are probably closely related, so they avoid mating with each other.

The scientists made the discovery by studying wild mice which were allowed to mate over several generations in a large outdoor enclosure. After 15 weeks the researchers studied the parentage of the animals to work out which mice had mated. They were surprised to find that the immune system genes previously

* *Current Biology*, vol. 17

believed to be the key to the process seemed to play no part whatsoever. 'Instead, another specialised set of proteins, which are produced at high concentration in mouse urine, signal relatedness through their scent. It is these proteins that allow animals to avoid mating with their close kin,' says Janet Hurst.

The key to the breakthrough was the use in the study of wild as opposed to laboratory-bred mice. Lab mice are all in-bred, so they carry the same urine proteins; as a result, researchers had focused instead on immune system genes.

GENETICS

CLONE ALONE, AND IT'S THE REAL 'MACAQUE-OY'

RESEARCHERS IN AMERICA have produced the world's first primate stem cell clones in a move that could ultimately see patient-specific stem cells on offer in the clinic. Oregon-based researcher Shoukhrat Mitalipov and his colleagues* successfully transferred the genetic material from the mature skin cells of a 9-year-old male rhesus monkey into egg cells collected from 14 donor female monkeys. The egg cells used in the experiment first had their own genetic material removed, and then the adult skin cell nucleus was delicately re-injected into the egg. After a short incubation period drugs were added to chemically kickstart cell division in the injected eggs. A small number (0.7%) subsequently began to divide, yielding embryonic stem cells (ES cells) that the researchers were able to grow in the dish.

The technique works because eggs contain a cocktail of chemicals that re-program DNA by switching on the correct combination of genes that are required for embryonic development. But despite huge effort, until now, scientists had not succeeded in making cloning

* *Nature*, vol. 450

technology work with primate cells – Dolly the sheep and Snuppy the dog were the furthest that they had progressed along the mammalian cloning path. Exactly why the technique should be so difficult in primate cells is a mystery, but Mitalipov and his team think that the key to their success was a more gentle approach to handling the egg. They avoided using ultraviolet light when removing the egg's own genetic material, and kept the concentrations of calcium and magnesium very low in the culture medium used to handle the cells.

The results, they say, prove that the technique can work in primates (and hence humans) and it should theoretically be possible in the future to produce stem cells tailor-made to an individual to repair damaged tissues and organs without the risk of immune rejection.

GENETICS

TWINS NOT SO IDENTICAL AFTER ALL

We might have to get used to the idea that identical twins are not quite the same after all, say researchers in the US. Harvard's Andrew Chess and his team, writing in *Science**, have found that of the two copies of each of our genes (one from our father and one from our mother), in at least 10% of cases one of the genes is randomly shut off. This means that in some tissues just the paternal copy of the gene is active, while in other tissues it's just the maternal gene. But the switch-off process isn't genetically determined – it takes place at random during development. So a stem cell makes a decision to deactivate one of the pair of genes, and then all of its progeny inherit that pattern. Alternatively, a different group of cells may leave both genes switched on. As a result, twins previously believed to be genetically identical, are almost certainly running subtly different 'genetic programs' throughout their bodies. They contain the same DNA as each other, but may well be switching on different bits of it.

What's the significance of the finding? Some diseases, such as cancers, are caused by the loss of function of

* *Science*, vol. 318, no. 5853, 2007

genes that stop cell proliferation. So if an individual already has one defective copy of one of these anti-cancer genes, but then deactivates the healthy copy in certain tissues, this could trigger tumour formation. Alternatively, failure to switch off one of the copies of a gene can lead to too much of that gene product, again triggering disease. An example of this is Alzheimer's disease, where overactivity of the beta-APP gene is linked to the formation of dementia-provoking protein plaques in the brain.

NON-HUMAN BIOLOGY

RESEARCHERS UNCOVER ACID-LOVING METHANE-MUNCHING BUGS

SCIENTISTS HAVE DISCOVERED one of the world's hardiest species of bacteria – methane-burning microbes that can tolerate temperatures of over 70° C and thrive in acid stronger than you'd find in a car battery. Known as *Verrucomicrobia*, the bacteria have been discovered by two independent research groups, one based in New Zealand*, the other in Europe.

The New Zealand-based team made their discovery at a site in Hell's Gate, a geothermal region of North Island. Examining the soil in a woody area that had recently been killed by escaping hot steam, the researchers measured the temperature, acidity and methane concentrations at different depths. They were surprised to find that at about 20 centimetres down, where the temperature was 70°C and the soil was reasonably acidic, methane levels suddenly nose-dived, suggesting that something at this depth was consuming the gas. From samples collected at this depth the team were able to culture a thriving colony of microbes that were eating methane. The bugs were pulling off their chemical stunt using an enzyme

* *Nature* vol. 450

called MMO – methane monooxygenase, which can convert methane to methanol; the bugs then convert the methanol to carbon dioxide, extracting energy in the process.

The European team uncovered a similar bacterium living in the equally harsh conditions of a fumarole near Naples. In both cases the microbes almost certainly help to sop up methane from the Earth's crust that would otherwise contribute to the atmospheric budget – and who knows, it might even be possible to press-gang these microbes into service to help us combat global warming by dealing with man-made methane.

MEDICAL TECHNOLOGY

HELPING HAND FOR AMPUTEES

SCIENTISTS IN CHICAGO are developing a superior control system for prosthetic arms to help users to make far more accurate movements – just by thinking. Todd Kuiken at the Rehabilitation Institute of Chicago* is pioneering an approach known as targeted muscle re-innervation (TMR). Patients are prepared for the device by first detaching nerves that supply the muscles of the chest wall that would once have helped to support and move their missing arm. Then, the motor nerves from the arm stump that would have supplied the limb are re-routed into the denervated chest wall muscles. After 6 months when the patients think about moving their missing hand or fingers the muscles of the chest respond. This can be picked up by an array of skin sensors and used to electronically control the movements of the prosthesis. Volunteers who have been equipped with the system have found that they can perform movements much more easily and intuitively than with conventional devices, which are activated using other muscles in the back. Now, to improve the system yet further, the researchers have

* *Journal of Neurophysiology*, vol. 98, 2007

developed computer software that studies the precise patterns of activity in the chest muscles and can take into account the timing and frequency of muscle movements. Using this approach they can discriminate, with 95% accuracy, between the electrical signatures of 16 different movements of the hand, arm and fingers. The next step will be to speed up the decoding process and link it to the prosthesis, so that users can use thought power to move their arm in realtime, something that the team predict they will be able to do within the next two years.

HUMAN BIOLOGY

EAT, SLEEP AND BE SKINNY

Researchers have shown that, somewhat surprisingly, the key to preventing children from becoming overweight is a good night's sleep. A growing body of evidence now links lack of sleep, increased appetite and weight gain amongst adults, but it wasn't clear whether the same applies to children as they grow up. A few studies have hinted at higher obesity rates amongst youngsters who take the least sleep, but it wasn't clear which effect came first, and whether being overweight just leads to poorer quality sleep.

To find out Julie Lumeng and her colleagues* at the University of Michigan enrolled nearly 800 9-year-olds and followed them up for 3 years, to see whether their sleep patterns (reported by their mothers) were related to changes in weight as they grew up. The results were astonishing. The children getting less than 9 hours shut-eye per night were much more likely to pile on the pounds over the next three years than their superiorly slumbering counterparts. Even more striking was the finding that every extra hour of sleep the children took at the age of 9 resulted in a 40% decrease

* *Pediatrics*, vol. 120, no. 5, November 2007

in obesity risk by the age of 12. And at age 12 each extra hour reduced the risk by 20%.

The researchers cannot say for certain why extra sleep should keep kids in trim, but there are several explanations. 'Children who are better rested may have more energy to get more exercise,' explains Lumeng. 'It's also possible that when children are tired they may be more irritable or moody and use food to regulate their mood.' It's likely that hormones also play a role. Amongst adults, sleep deprivation has been linked to a reduction in levels of the appetite-suppressing hormone leptin, and increased production of the appetite-boosting hormone ghrelin. These changes translate into a stronger urge to eat, increased calorie intake and weight gain. So, paradoxically, kids might be able to justify missing the school bus in future on the grounds that they're battling the bulge . . . with a lie-in!

HUMAN BIOLOGY

STEM CELLS RESTORE MISSING MEMORY

SCIENTISTS HAVE DISCOVERED that forgetful mice with damage to their hippocampus – the part of the brain that lays down new memories – can have their intellects restored by injecting neural stem cells collected from newborn mice. UC Irvine researcher Frank LaFerla and his team genetically programmed mice to develop degeneration of hippocampal nerve cells. In tests designed to measure recall of objects and an animal's surroundings, the engineered mice were significantly impaired; they showed signs of recognising their environment only 40% of the time, compared with 80% when control animals were tested. Next, the researchers injected into the hippocampi of some of the impaired animals 200,000 neural stem cells collected from the brains of newborn mice. The injected cells were programmed to produce a glowing green jellyfish protein so that the researchers could follow where the cells went in the brain.

The animals were then regularly re-tested to look for signs of recovery. After 3 months mice that had received stem cell transplants were remembering their surroundings 70% of the time – almost as good as control animals. In contrast, the untransplanted mice still showed memory deficits. To find out how the improvements had occurred, the researchers then studied the brains of the transplanted animals.

They were surprised to see that only about 4% of the cells had turned into neurones, indicating that the improvement wasn't being achieved solely through replacement of missing nerve cells. Instead, the researchers think that the stem cells stimulate repair and re-wiring of the nervous system by releasing beneficial chemicals such as growth factors.

But irrespective of how it works, as Frank LaFerla points out, 'Our research provides clear evidence that stem cells can reverse memory loss. This gives us hope that stem cells someday could help restore brain function in humans suffering from a wide range of diseases and injuries that impair memory formation.'

HUMAN BIOLOGY

'BREAST'-FOOD FOR THOUGHT

SCIENTISTS HAVE UNCOVERED the mechanism by which breast-feeding can boost brainpower. It's not all in the mind so much as all in the genome, because researcher Terrie Moffitt* studied over 3000 children from Britain and New Zealand and identified two forms of a gene called FADS2 that is carried on chromosome 11 and helps to process fatty acids. Carriers of the 'C' form of the gene, which is carried by 90% of the population, gain a 7-point IQ benefit when they are breastfed, presumably because their developing brains can make the most efficient use of the arachidonic and docosahexaenoic acids that are present at high levels in breast milk.

This means, says Moffitt, that if every baby was breastfed there would potentially be twice as many gifted children. But most significantly, the result lends biochemical credence to studies going back many years that have consistently shown that breastfed babies are at an IQ advantage.

---

* *PNAS*, vol. 104, no. 47, 20 November 2007

FANCY TRAINERS MAKE TAKE YOU DOWN THE ROAD TO RUIN

A NEW STUDY in the *British Journal of Sports Medicine* compared the relative merits of nine pairs of trainers ranging in price from 'economical' to 'ludicrously expensive' and found no evidence that the size of the price tag makes any difference to comfort or 'bone benefit'. Richard Clinghan and his colleagues* at Dundee University used pressure-measuring equipment to monitor how well the trainers compensated for the loads applied to the bones and joints of the foot under gentle and vigorous exercise conditions, and asked wearers to subjectively rate their comfort. The volunteers in their study were not told the price of the footwear, and any branding was hidden with sticky tape. At the end of the study there was no clear relationship between price and comfort, and the authors conclude that cheap trainers offer at least as good, if not better, cushioning to the bones of the feet compared with more expensive, flashier footwear.

---

* *British Journal of Sports Medicine*, published online, 11 October 2007

PRAWNS FEEL PAIN

NEXT TIME YOU drop a marine crustacean into a pot of boiling water you should feel a twinge of guilt, because new evidence suggests that the creatures probably do find the experience decidedly uncomfortable.

Queens University Belfast researcher Robert Elwood and his colleagues* dabbed acetic acid (vinegar – how appropriate!) on one of the two antennae of 144 prawns. The animals then spent the next 5 minutes rubbing the affected antenna in the same way that a mammal rubs a sore patch of skin, evidence, say the researchers, that these animals feel pain. 'The prolonged, specifically directed rubbing and grooming is consistent with an interpretation of pain experience,' says Elwood.

Some critics are unconvinced, arguing that the animals, which lack a brain, were just trying to clean the affected antenna; others have also pointed out that even single-celled organisms like bacteria can sense a threatening chemical gradient and move away from it, but not because it's painful. Elwood points out that such comparisons are wrong. 'Using the same analogy, one could argue crabs do not have vision because they lack the visual centres of humans,' he says.

---

* *Animal Behaviour*, vol. 1016, July 2007

But if it does turn out to be true then the discovery adds credence to the old adage 'no pain, no gain', because prawns taste great!

ASTRONOMY

THE PLANETS SUITE

SPACE SCIENTISTS HAVE discovered a fifth planet in a nearby star system, making it the largest planetary group yet observed around a distant star. Astronomers have been watching the distant star, which is called 55 Cancri and is 41 light years (or 5.8 trillion miles) away in the direction of the constellation of Cancer, for the past 18 years.

The system is too far away to see individual planets, but they can be detected by picking up telltale wobbles in the light coming from the star caused by the gravitational pull of orbiting bodies. Using this technique, in the mid-1990s, researchers found a planet the size of Jupiter orbiting the star every 14 days. Then, between 2002 and 2004, a further three planets were discovered. One, which is similar in size to Neptune and about 14 times bigger than the Earth, sits sizzlingly close to the star and completes an orbit in just 2.8 days. The other two planets include a massive body, which is four times larger than our own Jupiter and orbits once every 14 years, and a Saturn-sized planet with a 44-day orbit.

The newly discovered planet is also huge and at least 45 times the size of Earth, but it's got scientists intrigued because it has a 260-day orbit, similar to the trajectory of Venus in our own solar system. This means that it's likely to be a warm world, like the Earth. And although it's much larger than the Earth, it may well have a number of rocky moons, and if these contain water then the warm temperatures may well make them wet and ideal places where life could flourish.

The discovery, which has been made by Debra Fischer from San Francisco State University and Geoff Marcy from the University of California at Berkeley*, marks the first time astronomers have found a planetary system with more than four planets. 'We now know that our Sun and its family of planets is not unusual,' says Marcy.

---

* *The Astrophysical Journal*, vol. 67, March 2008

MEDICINE

VIRUSES VANQUISHED WITH A BURST FROM A LASER

SCIENTISTS IN THE US have developed a laser technique to selectively wipe out viruses and bacteria while leaving human cells unharmed. The approach could be used to improve the safety of blood transfusions and other blood-derived products without the risk of harming the cells present. Kong-Thon Tsen and his colleagues at Arizona State University* made the discovery when they exposed solutions of viruses including bacteriophages and TMV, a plant virus, to a train of very short (femtosecond duration) bursts from a near-infrared (850nm) laser. After an hour the viruses were completely deactivated. The team then repeated the experiment using *E. coli* bacteria and, again, after one hour the bacteria were all destroyed. But when they exposed solutions of human cells to the same treatment, the cells remained viable and healthy.

The researchers think that the laser pulses rip apart the outer coats of the microbes in the same way that it's possible to smash a glass with sound waves. The bursts of laser light cause a phenomenon called impulsive stimulated Raman scattering (ISRS), which sets

---

* *Journal of Physics: Condensed Matter*, vol. 19

up patterns of lethal vibrations amongst the proteins on the outsides of the micro-organisms, breaking them open. But mammalian cells, possibly because they are much larger, require between 20 and 100 times more energy to trigger the same process, so they're relatively invulnerable to the effect. As a result the team are confident that the technique, which should also be immune to bugs becoming resistant to its effects, could play a major role in the clinical setting.

'Although it is not clear at the moment why there is a large difference in laser intensity for inactivation between human cells and micro-organisms such as bacteria and viruses, the research so far suggests that ISRS will be ready for use in disinfection and could provide treatments against the worst, often drug-resistant, bacterial and viral pathogens,' says Tsen.

COLD SORES BAD FOR BRAIN AS WELL AS LOOKS

UK SCIENTISTS HAVE discovered that the herpes simplex virus (HSV), which is responsible for triggering cold sores, may also be linked to Alzheimer's disease. Professor Ruth Itzhaki and her colleagues* at Manchester University found that when brain cells grown in culture were infected with the virus they dramatically increased their production of a protein called beta-amyloid 1-42. This protein is known to accumulate, forming pathological deposits known called plaques, in the brains of Alzheimer's patients.

There are now a number of lines of evidence linking HSV infection with senile dementia. These include the discovery of viral DNA in the brains of patients who have died of the disease, the finding that one of the viral coat proteins is very similar to the structure of beta-amyloid, which might encourage the proteins to aggregate into plaques, and a form of a gene called ApoE4 is known to be a risk factor for both Alzheimer's and herpes infection.

Itzhaki and her team now suspect that a virus dormant in the brain periodically reawakens, triggering

---

* *Neuroscience Letters*, vol. 429

increases in beta-amyloid production and the progression of plaque formation. But there is some good news. On the basis of their present findings the team suggest that there are strong grounds for putting some patients with Alzheimer's on anti-herpes medication to ward off the progression of dementia. 'I think there's enough evidence to begin giving people with ApoE4 antiviral drugs to slow the progression of the disease,' says Itzhaki.

ECOLOGY

PLASTIC POISONS WORMING THEIR WAY UP THE FOOD CHAIN

RESEARCHERS AT PLYMOUTH University in the UK have uncovered worrying evidence of how, thanks to plastic waste in the sea, organic poisons make their way up the seafood chain and ultimately into us. Scientists have known for some time that plastic washed into the sea breaks up into tiny particles, called microplastics, which can absorb oil-based pollutants such as PCBs, nonylphenol and DDT, which are present at low levels in the water. They also suspected that when these particles settle on the seafloor, sediment dwellers like marine worms would ingest them and concentrate their chemical cargo of organic pollutants. Then, when the worms fell victim to predators such as fish and crabs, the chemicals they contained would pass up the food chain.

Now Emma Teuten and her colleagues* have proved that this is indeed the case. Writing in *Environmental Science and Technology*, the researchers report that they tested the effects of adding small quantities of plastic mixed with the organic pollutant phenanthrene to tanks containing sediments inhabited by

* *Environmental Science & Technology*, vol. 41, no. 22, 2007

lugworms (*Arenicola marina*). Compared with control worms, these plastic-exposed animals were predicted to be carrying 80% higher tissue concentrations of the pollutant. 'It [plastic] accelerates the mechanism by which contaminants accumulate in the food chain,' says Teuten. The team suggest that more research is needed to understand the implications of their findings, including looking at other types of plastics, and how they affect the persistence of pollutants in the environment.

PLANT BIOLOGY

EDELWEISS, MUSIC TO THE EARS WHEN IT COMES TO SUNSCREEN

RESEARCHERS STUDYING THE high Alpine plant edelweiss may have stumbled on an impressive new sunscreen strategy. Brussels-based researcher Jean Pol Vigneron and his colleagues* wondered how the edelweiss plant was able to withstand so effectively the high doses of UV radiation that accompany life at high altitude.

To find out, they measured how the plant absorbed or reflected lights of different wavelengths, expecting to see the UV being bounced back at them. But they were shocked to see that while most wavelengths were reflected, UV wasn't. 'It's astonishing, but the plant completely absorbs the UV,' says Vigneron. Edelweiss leaves are covered in fine white hairs and the team wondered whether they might be responsible for soaking up the UV.

They studied them under the electron microscope, which revealed that the hairs are made up of tiny fibres, each just 176 nanometres (millionths of a millimetre) across. This diameter means that the fibres are about the same size as the wavelength of ultraviolet light,

* *Physical Review E Statistics Nonlinear Soft Matter Physics*, vol. 71, January 2005

and can interact with it.

The researchers think that the tiny strands steer the UV away from the leaf where it could do harm and into the middle of the hair where it can be soaked up by a material in the centre, possibly pure water, which is a good absorber of UV.

Vigneron suspects that the same trick could also work for sun-lovers seeking better protection than that offered by standard titanium-based creams and lotions currently on the market.

VIROLOGY

SCIENTISTS DISCOVER BACTERIAL CELL SUICIDE SWITCH

SCIENTISTS IN ISRAEL have uncovered a chemical suicide signalling pathway that turns *E. coli* into the bacterial equivalent of lemmings, a discovery which could lead to a new generation of powerful antibiotics. Writing in *Science*, Ilana Kolodkin-Gal and her colleagues* at the Hebrew University in Jerusalem describe the discovery of a 'death factor', which triggers *E. coli* to kill themselves. *E. coli* carry a gene 'module' called mazEF, which produces a long-lived toxin called MazF and a short-lived anti-toxin MazE. So long as the genes remain switched on all is well, because the antitoxin blocks the action of the toxin. But if the cell is stressed – such as might occur when bacteria are attacked by viruses, or if food becomes scarce – the gene switches off and the short-lived antitoxin disappears. This leaves the stable MazF unchallenged to kill the bacterium, which it does by chewing up genetic material inside the cell.

The bugs activate this system when their population exceeds a certain level, the purpose being to ensure the survival of other members of their colony, because

* *Science*, vol. 318, 2007

self-sacrifice can be used to deal with virus infections or lack of nutrients.

However, the chemical identity of the molecule known as EDF that trips this suicide switch was not known. To find it, the team collected extracts from the media used to grow dense cultures of bacteria, and then picked through the chemicals individually to identify the EDF molecule, which turns out to be a short protein sequence. The researchers were then able to make an artificial version of the protein that could fool small numbers of *E. coli* into thinking they were part of a much larger population, which primed their mazEF cell suicide programme. When the bugs were then stressed with a small dose of an antibiotic, they obliged by killing themselves.

Coming at a time when the world is fast running out of antibiotic options, the discovery of this pathway could prove critical in the discovery of new ways to combat the rise of the superbug.

GENETICS

AS WELL AS RED-BLOODED, NEANDERTHALS WERE RED-HEADED

A TRAWL THROUGH the emerging Neanderthal genome has revealed that our hominid relatives probably had red-heads amongst their numbers. Writing in *Science*, Carles Lalueza-Fox from the University of Barcelona, together with an international team of collaborators*, has successfully sequenced the melanocortin 1 receptor gene (mc1r) from two Neanderthal specimens, one Spanish and one Italian.

The mc1r gene controls how cells respond to a hormone called MSH – melanocyte stimulating hormone. It's known that among modern humans variations in this gene affect the amounts of the two pigment molecules eumelanin and phaeomelanin that are made in the skin, with red-heads and pale-skinned individuals making proportionally less eumelanin.

When the researchers used the polymerase chain reaction to copy part of this gene from the Neanderthal remains they found a variant of the gene never before seen among modern humans. To find out what effect it might have they added the gene to cells grown in the dish and then treated the cells with MSH. The cells

---

* *Science*, vol. 318, no. 5855, 30 November 2007

behaved almost identically to cells containing the mc1r gene from modern pale-skinned humans, suggesting that the Neanderthals who originally carried this gene variant would also have had a pale complexion and red hair. The team estimate that, on the basis of their findings, at least 1% of the Neanderthal population probably looked like this.

The result is also interesting because it adds weight to the idea that pale skin suited populations living north of the Equator, and also seems to disprove the claim that Neanderthals disappeared because they interbred with modern humans. If this were true, then we should see this red-headed Neanderthal mc1r variant cropping up in modern humans, but so far that hasn't happened.

BROCCOLI COMBATS CANCER-CAUSING UV DAMAGE

PRESIDENT BUSH SNR might have refused to eat it, but if he wants to ward off skin cancer perhaps he should give broccoli the benefit of the doubt, because new research suggests that it powerfully mitigates against damage done to the skin by UV radiation. Writing in *PNAS*, Johns Hopkins researcher Paul Talalay and his colleagues* applied a broccoli extract to small patches of skin on the backs of six volunteers over a 3-day period. The subjects were all asked to avoid broccoli and other cruciferous vegetables for a week before the study. The treated area, together with a similar-sized untreated area used as a control, was then irradiated with a dose of ultraviolet light and a chromometer, a colour-sensitive device, was used to measure the redness of the skin afterwards. The team found that the broccoli-treated patches of skin were, on average, 37% less burned (judging by their degree of redness) than the untreated patches of skin.

The team have previously found that a component of broccoli, called SF – sulforaphane – can trigger the production of an effect known as a 'phase 2 response'

* *PNAS*, vol. 104, no. 44, 30 October 2007

in treated cells. This involves switching on genes such as NQO1, GSTA1 and haem oxygenase 1, which are known to protect cells from damage produced by agents like UV. 'This is the first demonstration that a human tissue can be protected directly against a known human carcinogen,' Paul Talalay said.

'But this is not a sunscreen,' he cautions, because it doesn't block UV, just enables cells to better handle its effects. So instead he thinks that the effect might be useful as an add-on in protecting people from UV, especially among people at increased risk of skin cancer, such as individuals who are immunosuppressed.

HUMAN BIOLOGY

VIBRATE YOUR WAY TO SVELTE-DOM

COUCH POTATOES EVERYWHERE will be rubbing their hands together when they hear that preventing weight gain need not involve a rigorous gym trip. Indeed, if a paper in *PNAS* has got it right, some of the vibrations from that hand-rubbing might just be all the exercise they need. That's because New York researcher Clinton Rubin and his colleagues* have found that a short daily stint on a vibrating platform resulted in mice that were much leaner than their less stimulated counterparts.

The researchers divided a group of similarly sized mice into two groups. One group were giving daily 15-minute vibration sessions. The platform on which they stood moved just 12 thousandths of a millimetre up and down 90 times per second. The movement was just perceptible to a finger laid upon it. Apart from not receiving vibration, the second group of control mice were otherwise treated identically to the first. After 15 weeks the results were striking. The vibrated mice had 23.5% less body fat than the controls and were 1.2 grams lighter, on average. CT scans of the animals' bodies also confirmed 27% less fat around their middles.

---

* *PNAS*, vol. 104, no. 45, 6 November 2007

To investigate where the fat was going, the team used a third group of mice in which their bone marrow had been genetically programmed to produce a green-coloured marker protein, enabling the researchers to track where cells produced by the bone marrow went in the body. Under normal circumstances at least a proportion of the cells turn into fat cells, called adipocytes. But when the animals were vibrated the number of fat cells being produced fell, suggesting that the stimulus was in some way altering the fate of these cells and causing them to turn into other tissues.

The team suggest that this could lead to a non-strenuous drug-free method for obesity control.

RESEARCHERS GET TO THE BOTTOM OF CORAL CLOCK

WORKING ON THE Great Barrier Reef, researchers at the University of Queensland in Australia have lifted the lid on one of the great marine mysteries of our time – how corals synchronise their spawning and ensure that it takes place on just a few nights of the year and always when there is a full moon. Writing in *Science*, Oren Levy and his colleagues* describe how they have found the coral equivalent of the human body clock in the larvae of reef-building coral (*Acropora millepora*). The team probed the coral DNA for sequences similar to genes called CRY – short for cryptochrome – which are used to power the body clocks, and in some cases directly detect light, in flies, worms and mammals.

Using this approach the team successfully identified two coral CRY genes, and then went on to measure the activity of the genes at different times of the day and night. They found that the genes encode proteins which respond to blue-green light, which fits perfectly because water absorbs red light very strongly. Then, by growing coral samples in tanks

---

* *Science*, vol. 318, no. 5849, 19 October 2007

with either a normal light-dark cycle to simulate day and night, or by keeping the coral just in the dark for an extended period, the team found that the levels of the genes peaked in the daylight and dropped off in the dark. Next they collected RNA samples from corals on the reef when there was a full moon, and again when there was a new moon (i.e. no moon) and compared the levels of the two genes. Intriguingly at the time of the full moon the levels of one of the genes – CRY2 – were much higher, suggesting that this gene is the linchpin which links the coral's behaviour to moonlight and therefore the spawning pattern.

'It's very complex and we don't yet know what the other factors are that regulate the process,' says Levy. 'For instance, it's one thing knowing that the time is right, but how does the coral know to spawn simultaneously with other corals? My guess is it will take another couple of years to understand how it all fits together.'

NON-HUMAN BIOLOGY

SCIENTISTS MUSSEL-IN ON STICKY SURFACES

US RESEARCHERS STUDYING how mussels anchor themselves to rocks, stones and jetties have stumbled upon a trick that can chemically coat any surface with something so sticky that it will even bind to Teflon. When Phillip Messersmith and his team* at Northwestern University in Illinois analysed the mussel glue they found it to be very rich in a chemical called DOPA (di-hydroxy phenylalanine), together with an amino acid building block called lysine. These two molecules interact to give the glue its sticking power.

So the researchers wondered whether other molecules, containing the same chemical groups, might behave in the same way when exposed to seawater. To find out, they used the substance dopamine, which is used in the body as a nerve transmitter but also contains both a catechol group, like DOPA, and an amino group, like lysine. Incredibly, the trick worked. When a small amount of dopamine was dissolved in water and then the water was made slightly alkaline like the sea, the dopamine molecules linked up to form polymers. If an object was added to the solution

---

* *Science*, vol. 318, no. 5849, 2007

as this process was taking place an ultrathin layer of the polymer just 50 nanometres thick (1000 times thinner than a human hair) was deposited on the surface, which then behaved as a 'key' to which other substances could be attached. By dipping treated objects into a solution containing copper or silver ions, it was possible to metal-plate the material, producing electrically conductive plastics. The team hope that the discovery will make it possible to produce body implants and medical implants, such as silver-coated catheters, which have anti-bacterial qualities.

It has even been possible to use the technique to remove heavy metal contamination from drinking water. 'We were able to remove mercury from water by passing it down a column containing beads treated with our polydopamine coating,' says Messersmith. 'So this could be very useful in cleaning up water in countries with heavy metal pollution'.

RNA-AWAY HEPATITIS

SCIENTISTS IN THE US have uncovered how the body tackles certain viral infections, and the results might help to produce better ways to treat hepatitis C. Hepatitis C is a blood-borne viral infection spread by needle-sharing, use of contaminated blood products and by sex, and it affects about 1% of the population in the Western world. Eighty per cent of individuals who catch it develop a chronic infection which causes liver injury and can lead to cirrhosis in about 20% of cases.

Recently doctors have found that long-term therapy with an immune-regulating hormone called alpha-interferon can enable the immune system to overcome the virus and eliminate it from the body, although exactly how remained a mystery. Now, writing in *Nature*, UC San Diego's Michael David and his colleagues* have uncovered the mechanism, and in the process revealed a previously unknown mechanism by which the body deals with viral infections.

The team treated cultured cells with interferon and then studied the levels of short pieces of genetic material

* *Nature*, vol. 449, 18 October 2007

called micro RNAs, which have in recent years been shown to play key roles in regulating gene activity. Using this technique the researchers pinpointed 30 different micro-RNAs that altered their levels in the presence of interferon. They then compared these RNA sequences with the hepatitis C genome to see if there were any matches. What they were looking for were sequences in the virus that were mirror images of the micro-RNA, because when this happens the micro-RNA can bind to its mirror image viral gene and disable it.

The researchers found eight micro-RNAs that were direct hits in the hepatitis C genome. To find out whether they could affect the growth of the virus they added synthetic versions of each of the micro-RNAs to hepatitis C infected cells, which reduced the turnover of the virus in the cells by 70%. This approach could hold the key to effective new therapies for hepatitis C in the future, and may also spare patients the unpleasant flu-like side effects of interferon therapy.

SOLAR TECHNOLOGY

WORLD'S SMALLEST SOLAR PANEL, OR SHOULD THAT BE WIRE?

US SCIENTISTS HAVE taken solar cell technology a step further with the development of the solar nanowire – a solar device 200 times thinner than a human hair. Charles Lieber and his team* at Harvard in the US have found a way to deposit the three semi-conductor layers needed to make a solar cell in concentric circles. The innermost layer consists of silicon containing a small amount of boron, which makes it eager to give up electrons. A thin insulating 'shell' of plain silicon is then deposited around the inner core, and an outer layer of silicon 'doped' with electron-hungry phosphorus is added around the outside. When light hits the wire it knocks electrons off the boron and into the phosphorus-containing outer shell. They then flow around a circuit to return to their starting point.

Under normal sunlight each wire pumps out about 50–200 pico-watts (less than a billionth of a watt), and they are only about 3% efficient, but the team are aiming for 15% efficiency and have already demonstrated that with light eight times the intensity of sunlight, they can power a nano-sized pH meter.

* *Nature*, vol. 449, 18 October 2007

'This enables us to study a fundamentally different geometry for photovoltaic cells,' says Lieber.

PILL BAD FOR LAPDANCERS' FORTUNES

A NEW STUDY by researchers at the University of New Mexico at Albuquerque and published in *Evolution and Human Behaviour* shows that female lapdancers seeking to supersize their earnings should eschew use of the oral contraceptive pill, because it makes them less attractive to tipping punters. Geoffrey Miller and his colleagues* recruited 18 lapdancers from local clubs and asked them to record their menstrual activity, earnings and contraceptive pill use over a 60-day period. The results were striking: the earnings of normally-cycling (non-pill-using) women peaked in line with their fertility, exceeding $350 per shift by mid-cycle. But their contraceptively compromised (pill-using) counterparts fared less well, earning an average of only $200 in tips per shift throughout their cycles. Significantly, both groups earned approximately the same amount during menstruation, indicating that the effect was not due to an overall difference in attractiveness between the women.

The results strongly argue the existence of a female 'oestrus', which prevailing wisdom claimed had been

* *Evolution & Human Behaviour*, vol. 28, Issue 6, November 2007

lost amongst humans during evolution. Instead, the results suggest, men can subconsciously detect when women are at their most fertile and tend to judge them as more attractive, valuable and, it would seem, worthy of larger tips at this time! So as well as health consequences, there may be an economic disadvantage to taking the pill too!

ANT USE-BY DATE

New research published in *Animal Behaviour* reveals that ants seem to be able to predict their own best-before dates. Dawid Moron* from Jagiellonian University in Poland artificially aged European *Myrmica scabrinodis* ants by exposing them to CO_2, which damages the nervous system, or by giving them small injuries to make them vulnerable to infection and dehydration.

Over the following five weeks the team then followed up how the injured ants fared against their uninjured nest-mates, and what roles they fulfilled in their colonies. Intriguingly, compared with uninjured ants of a similar age, the compromised individuals tended to take on higher risk tasks, such as foraging outside the nest, at a much younger age than normal. And the more severe their injuries, the sooner they took on these tasks. Such roles are usually assigned only to much older ants, which are deemed more expendable because they are closer to the ends of their lives. They are also more likely to carry parasites and other infections that might pose a threat to the colony, so it makes sense for them to spend more time outside the

* *Animal Behaviour*, vol. 75, Issue 2, February 2008

nest, compared with more valuable younger healthier specimens, who are confined to barracks initially.

These new findings shed light on how this pecking order might be achieved, which is at least partly due to ants having a strong grasp of their own mortality, and even their 'time to live'!

MOOSE 'BEAR' PROTECTIVE EFFECTS OF ROADS IN MIND

RESEARCHERS IN AMERICA'S Grand Teton National Park have found that moose have learned to use roads to shield their calves from grizzly bear attacks. Writing in *Biology Letters*, Joel Berger and his team* from the US Wildlife Conservation Society describe how they began radio-tagging 18–25 female moose annually in 1995. They used the tags to monitor the behaviour and movements of the animals, including where and when they gave birth to young. As time went on the researchers began to notice a pattern emerging. The animals were moving closer and closer to roads to drop their calves, in some cases as near as 50–100 metres.

Now Berger thinks that rising numbers of grizzly bears are the cause. Studies in Alaska show that grizzlies account for 90% of deaths among young moose, but the bears won't go near roads or other signs of human activity, typically remaining over 500 metres away. Female moose, on the other hand, tend to avoid locations where there are bear droppings, particularly if they have lost a calf to a bear in the past. As a result, the moose are now seeking sanctuary closer to roads.

* *Biology Letters*, vol. 3

Support for the theory comes from the observation that only animals in the north of the park seem to be showing a preference for using a nearby road as a nursery – and Yellowstone Park, where grizzlies were granted protected status in 1993 and are now present in large numbers, is only a few tens of kilometres away further north.

But the protection of the roadside may be short-lived because there are now signs that the bears are becoming more audacious. 'So far the bears have avoided these areas,' says Berger. 'But if you're a hungry bear why wouldn't you go towards the roads?'

HUMAN BIOLOGY

RESEARCHERS HIT PAIN WHERE IT HURTS

THE ABILITY TO selectively switch off pain while leaving other nerve pathways unaffected has been a holy grail of anaesthesia for decades. Now researchers in the US have cracked the problem with the help of a redundant anaesthetic agent and the juice of a chilli. Writing in *Nature*, Alexander Binshtok and his colleagues* from Harvard Medical School have found that the combination of capsaicin, the chemical that makes chillis hot, together with a local anaesthetic derivative called QX-314 that doesn't normally penetrate nerve cells, can render rats insensitive only to pain. The team injected the spicy anaesthetic mixture into the animals' hind limbs, adjacent to their sciatic nerves. The animals continued to walk around normally, but showed a dramatically reduced sensitivity to painful and thermal stimuli.

Neither agent injected on its own reproduced the effect, so what was the basis of this painkilling combination? It turns out that QX-314 does not normally work as a local anaesthetic because the molecule has a permanent positive charge. This prevents it from

* *Nature*, vol. 449, 4 October 2007

penetrating the oily membranes of nerve cells, which it needs to do in order to block pores in the membrane that are essential for nerve excitation, and this can only occur from inside the cell. But adding capsaicin provides the anaesthetic agent with an alternative route into the cell by opening a channel in the membrane called TRPV1. So when the capsaicin prises open the channel, the QX-314 slips inside. And because TRPV1 is only found on nerve cells that convey pain signals, only pain-bearing fibres are inactivated.

The team are confident that the discovery will make a major impact on surgical and post-op anaesthesia. 'Eventually the method could completely transform analgesia, allowing patients to remain fully alert without experiencing pain or paralysis,' says co-discoverer Professor Clifford Woolf.

ECOLOGY

AGE NO IMPEDIMENT TO HOT SEX

SCIENTISTS IN THE US and Australia have shown that age is not a barrier to hot sex, at least when it comes to one of the world's oldest species of tree. Reporting their findings in the journal *Science*, University of Utah researcher Irene Terry, together with Gimme Walter, Chris Moore and Craig Hull* from the University of Queensland, have solved a longstanding puzzle about the reproduction of cycads, primitive plants dubbed 'living fossils' that date back almost 300 million years.

Like us, the plants come in male or female forms and both produce cones similar to those seen on modern fir trees. Pollen from the male cones must make its way to the female cones to fertilise them, and previously scientists thought that the wind was responsible, but subsequent studies showed that the cone scales were packed too tightly for pollen to enter efficiently.

Instead, it turns out that the cycads rely on thrips, tiny insects that feast on pollen, to get the job done. But how do you persuade pollen-covered insects to

* *Science*, vol. 318, no. 5847, 5 October 2007

vacate a male cone and head for the female counterpart? The answer is ingenious. The cycad cones lure thrips by secreting low concentrations of volatile chemicals, including one called beta-myrcene, which the insects find attractive. The male cones produce more of the chemical than the female cones and so tend to attract more insects initially. But around midday the cones abruptly warm up, sometimes by as much as 12° Celsius, and remain hot until mid-afternoon, a feat they achieve by burning up sugars, starches and fats stockpiled within the cone.

The surge in temperature triggers a huge million-fold increase in the production of the volatile chemicals, chiefly in the male cones. 'It takes your breath away. It's a harsh, overwhelming odour like nothing you've ever smelled before,' says Terry. Just as it repels humans, the smell also overwhelms the thrips, which desert the male cones in droves, still covered in the pollen-vestiges of their dinner. Later, as the day goes on and the cones cool, the odour concentration drops and the insects find it attractive again and return. But they also visit some of the female cones, mistaking them for males, and in the process deposit some of the pollen they picked up earlier. The cycle repeats itself the next day 'until the males wear out and the females are happily pollinated', says Terry.

So 300 million years old they might be, but they're not so unlike humans after all, it seems!

NON-HUMAN BIOLOGY

KNOWLEDGE OF WHERE HATCHLINGS GO TURNED TURTLE

SCIENTISTS HAVE SOLVED a longstanding mystery of the sea – what happens to newly hatched turtles after they dig themselves out of their sandy incubator and drag themselves down to the water? Previously the extent of our knowledge ended at the moment the animals entered the ocean, but now chemistry has come to the rescue.

In a paper in *Biology Letters*, Kimberly Reich and colleagues* from the University of Florida describe how they have used chemical isotopes to track down what happens to newborn green turtles. The team collected samples of the animals' shells and analysed the different isotopes they contained. Isotopes are heavier or lighter forms of the same chemical element, and certain processes preferentially use one of those forms, so this chemical fingerprint can be used to trace an animal's diet or where it has been living.

The researchers found that the oldest parts of the turtles' shells were very rich in the isotope nitrogen-15, which tends to be concentrated up the food chain, indicating that the animals were spending their first

* *Biology Letters*, vol. 3

3–5 years as carnivores. The same regions of the turtle shell also contained very low levels of carbon-13, which proves that the animals were living in the open ocean, because carbon-13 tends to be at higher concentrations in shallow water where plants snap up the lighter carbon-12 form for photosynthesis.

RESEARCHERS PAINT A PICTURE OF VOLCANOLOGY

SCIENTISTS HAVE RESORTED to an unusual source of evidence to trace the history of volcanic eruptions on Earth and their effect on the climate – by studying paintings. The effects of historical eruptions on the composition of the atmosphere are difficult to estimate owing to a lack of measurements or samples surviving from those times.

But now Christos Zerefos from the Academy of Athens in Greece and his colleagues* have come up with a proxy measure that could be used to shed light on the problem – old masterpieces. They analysed the sunsets immortalised in hundreds of paintings from the last 500 years, including works by Turner, Klimt and Degas. Using digital images of the paintings, the team calculated the red/green colour ratios and found that artists used much more red after major volcanic eruptions such as the explosions at Krakatoa in 1883 and 1680. They then used the red/green ratios from the paintings to calculate how much light was being interrupted by volcanic ash particles in the air. Their results tally well with what we know of the composition of the

* *Atmospheric Chemistry and Physics*, vol. 7, 2007

atmosphere at these times, suggesting that this could be an effective way to historically reconstruct the stratosphere.

CAFFEINE MONITOR

Scientists have come up with a sensitive new way to analyse the composition of almost any surface, including human skin. Renato Zenobi and Huanwen Chen* from ETH Zurich in Switzerland blow a stream of nitrogen across the surface under analysis and collect the gas together with any debris that it dislodges. The material is then fed into a mass spectrometer that can pick apart the chemical composition of anything present on the surface.

Using the technique, say the researchers, they can even detect the difference in caffeine concentrations in an individual before and after a cup of coffee. This means that the approach could be used to monitor drug levels, treatment compliance and address other important questions such as dissecting the chemical make-up of skin samples, plant tissue, frozen meat or other tissues. Critically the process is entirely non-invasive and does not harm the surface under analysis.

So caffeine-heads, watch out, someone could be monitoring your dose quite soon!

---

* *Angewandte Chemie International Edition*, vol. 46, Issue 40

SCIENTISTS UNCOVER KEY TO CANCER SPREAD

Scientists have discovered a molecular switch that turns on a cancer cell's ability to spread to other parts of the body. Publishing in *Nature*, MIT researcher Robert Weinberg and his colleagues* were examining micro-RNAs, small pieces of single-stranded genetic material produced in the cell nucleus. The actions of these RNA sequences are still poorly understood, but they are known to alter the activity of other genes. So, to find out how they might affect the behaviour of certain cancers, the team set out to compare the levels of 29 different micro-RNAs in tumour cells and healthy tissue. The team also looked at how the levels varied between cancers that had already begun to spread (metastasise), and those that hadn't. Intriguingly, one of the micro-RNAs – called micro-RNA-10b – was present at much higher levels in the metastatic tissue, suggesting that it might be involved in triggering the process. To find out, the researchers increased the levels of micro-RNA-10b in some human breast cancer cells and implanted them into mice. Compared with animals injected with the unmodified cells, animals

* *Nature*, vol. 449, 11 October 2007

that received cells containing higher levels of micro-RNA-10b rapidly developed spreading cancers. Next, to find out how micro-RNA-10b was triggering this process the team used a computer program to screen for 'targets' – other genes – that might be affected by the micro-RNA. They were able to home in on a gene called HoxD10, which works like a cellular handbrake, preventing cells from going AWOL. It's switched off by micro-RNA-10b during embryonic development so that cells can migrate to their correct future locations in the developing body, but in adulthood it is strongly expressed and helps to keep cells stationary.

'During normal development, this micro-RNA probably enables cells to move from one part of the embryo to another,' points out Weinberg. 'Its original function has been co-opted by carcinoma [cancer] cells.' Excitingly, when the researchers increased the levels of HoxD10 in experimental cancers the cells lost their ability to migrate and invade. This means that it may be possible to exploit the discovery as a powerful anti-cancer target: 'I was able to fully reverse micro-RNA-10b–induced migration and invasion, suggesting that HoxD10 is indeed a functional target', says Li Ma, lead author in the study.

DISEASES AND DISORDERS

'CATCH-TISH-FLU' – RESEARCHERS DEVELOP BETTER FLU-DETECTING TOOL

SCIENTISTS AT THE Institute of Bioengineering and Nanotechnology in Singapore, writing in *Nature Medicine**, have developed a novel technique that uses tiny magnetic silica-coated particles to purify and then diagnostically copy flu virus genetic material from a patient's throat swab.

The entire process takes place inside a tiny oil-covered water-droplet, which is prepared from the patient's sample and placed on a teflon-coated glass slide. The oil prevents the water in the droplet, which contains just one-tenth of a millilitre of liquid, from evaporating. Once the droplet is in situ the particles are added by stirring them into the droplet using a permanent magnet placed beneath the glass. As the magnetic particles are stirred in the solution, they soak up any viral RNA (the genetic material of flu) and are then withdrawn from the droplet. Next they are passed across the glass slide, again using the magnet, through a series of smaller droplets containing a cleansing solution. This washes

---

* *Nature Medicine*, vol. 13, 2007

away impurities clinging to the flu RNA or the particles. After four washes, the particles are dragged into a final droplet containing a chemical to release the RNA from the particles, and the ingredients for PCR (polymerase chain reaction), a process which copies or amplifies genetic material. Once sufficient genetic material has been produced it can be analysed to pinpoint whether it contains the signature of H5N1.

Juergen Pipper, the lead author on the paper, points out that this approach yields 50,000% more RNA, is 50 times cheaper and four times faster than existing diagnostic methods, which might make a life or death difference when it comes to controlling the next pandemic.

DISEASES AND DISORDERS

ANOREXIA LEAVES BAD TASTE IN THE MOUTH

SCIENTISTS HAVE DISCOVERED that the brains of anorexics respond differently to certain tastes than the brains of control subjects, possibly explaining why sufferers eschew tasty foods. Writing in *Neuropsychopharmacology*, University of Pittsburgh researchers Walter Kaye, Angela Wagner and their colleagues* brain-scanned 16 recovered anorexics as they were fed a sweet drink containing 10% sucrose, or just plain water, and compared their patterns of neural activity with those detected in 16 healthy women.

Among the normal subjects, whenever the sugary stimulus was presented, a region of the brain's grey matter known as the insula cortex lit up. The increase in activity in this region also tallied with the subjects' reports of how pleasant they found the sweet liquid to be.

But among the recovered anorexics the levels of activity detected in the insula were much lower in response to both the plain water and sugar solutions. The finding fits in with previous studies on this part of the brain, which have shown that the insula seems

* *Neuropsychopharmacology*, vol. 33, no. 3, 2008

to be involved in processing how the 'value' of certain foods might affect the body. For instance, animals with damage to this brain region lose the ability to avoid foods that have previously made them sick, and fail to stop eating high-calorie foods when they are full. This suggests that the insula may help to translate the experience of eating foods into pleasurable sensations, but this function seems to be abnormal among people with anorexia, possibly explaining why they avoid 'pleasurable' foods, fail to respond appropriately to hunger and lose so much weight. 'We know that the insula and the connected regions are thought to play an important role in interoceptive information, which determines how the individual senses the condition of the entire body,' says Kaye.

DELHI BELLY SENT PACKING

SCIENTISTS HAVE DECODED the genome of the common gut parasite *Giardia lamblia*, a frequent cause of gastric upset amongst backpackers and even preschool children. The genome sequence, published in *Science* by Mitchell Sogin and his colleagues* from the Marine Biological Laboratory at Woods Hole, US, should help scientists to develop better ways to prevent and treat the infection.

This is because *Giardia* leads a doppelganger life-cycle: it spends part of its life swimming around inside the intestines as a microscopic 'trophozoite', soaking up nutrients and multiplying. Then, when it is shed from the body, it forms a tough cyst that can survive harsh chemical and environmental conditions. But, when an individual ingests one of the cysts, usually by drinking contaminated water, the acidic environment of the stomach resuscitates the dormant cyst so that, within hours, it has turned into an intestinal troph-ozoite. To be able to coordinate such a complex life-cycle, the bug requires a significant genetic arsenal, which can also allow it to bypass the action of many

* *Science*, vol. 317, no. 5846, 28 September 2007

of the drugs used to control it. 'Existing drugs can effectively treat people with *Giardia* infections but, as with many pathogens, the concern is that the parasite will develop resistance to these medications,' says Anthony Fauci, from the National Institute of Allergy and Infectious Disease, which funded the study. 'But the *Giardia* genome shows us that the parasite has a large complement of unusual proteins that are potential targets for new drugs or vaccines.'

ECO-IDOL: HINDU FESTIVALS THREATEN INDIAN RIVERS

As each festival season approaches for Indian Hindus, environmentalists try to avoid a catastrophe caused by the faithful tossing millions of decorated statues of gods into rivers and waterways.

The usually elaborately decorated effigies often contain toxic levels of heavy metals, including lead, mercury and chromium, as well as cancer-causing dyes and plaster compounds that can deplete water oxygen levels. This leads to severe water pollution, which can kill fish and other aquatic organisms and threaten human health. 'The commercialisation of holy festivals like Ganesh Chaturthi and Durga Puja has meant people want bigger and brighter idols and are no longer happy with the ones made from eco-friendly materials,' said Ramapati Kumar, a toxics campaigner for Greenpeace India. 'Traditionally, the idols were made from mud and clay and vegetable-based dyes were used to paint them, but now it's more like a competition between households and between corporates who sponsor the idols to gain publicity.' About 80% of India's 1.1 billion population are Hindus, but in recent years their religious activities have been subject to increased scrutiny due to growing public awareness of environmental issues. 'No one is saying the immersion of idols should not happen,' says the Centre for Science and Environment's Suresh Babu*, 'but the government should impose guidelines to craftsmen who make the idols to use eco-friendly materials and organic paints so that we give the environment as much respect as we give God.'

* www.enn.com/lifestyle/

HUMAN BIOLOGY

'BRAIN-CLOTTING' – NEW MOVIE REVEALS ORIGIN OF PLATELETS

A Harvard-based research team have successfully produced a miniature movie of the generation of platelets, the key elements that allow blood to clot. Tobias Junt and his co-workers* used a fluorescent dye to label platelet-producing cells in the bone-marrow of mice and then watched in real time as the cells threaded thin extensions of their membranes into nearby blood vessels. Once *in situ* the current of passing blood caused the thin, finger-like projections to fragment, producing platelets. Although the origin of platelets was previously known, the steps involved in their production were not. So the improved understanding brought about by these amazing movies, which are reported in *Science*, may help in the management of bleeding disorders and other conditions associated with low levels of circulating platelets.

---

* *Science*, vol. 317, no. 5845, 21 September 2007

ANY ADVANCE ON HIV?

SCIENTISTS HAVE TAKEN a step forward in working out what's required to make a vaccine to tackle the world's worst pandemic – AIDS. Dennis Burton and Anne Hessell*, from the Scripps Institute in California, have shown that injecting an antibody called b12 can protect macaques against infection with the monkey equivalent of HIV. But when the team, who have described their findings in *Nature*, modified certain parts of the antibody they found that, despite still being able to recognise the virus, the antibody could no longer effectively protect experimental animals. The critical region seems to be the way the antibody recruits other components of the immune system, including white blood cells and a pathogen-disabling family of proteins called complement. What this study shows is that it should be possible to produce a vaccine to stop HIV, but it will be critical to ensure that it drives the production of antibodies that can harness other arms of the immune system like the b12 antibody used in this study.

---

* *Nature*, vol. 449, 2007

SMELLS GOOD TO ME

RESEARCHERS HAVE DISCOVERED the first evidence that our experience of smell is not general but highly personal and that genes govern whether an odour smells nice, or nasty.

Writing in *Nature*, Leslie Vosshall and her colleagues* at Rockefeller and Duke Universities say they asked human volunteers to smell androstenone, which is a hormone derived from testosterone and produced in large amounts by males. Intriguingly, the volunteers' experiences of the chemical were wildly different. Some described it as 'sweet or floral', others found it odourless and thought they were being tricked or asked to smell water, while a large number were repulsed, describing the smell as 'sweaty or urinous'; as Vosshall herself says, 'To me it smelled like the armpit of a man who had run a hundred miles without taking a shower.'

The team then genetically screened the volunteers and found a different version of a smell-receptor gene called OR7D4 in those who experienced the smell differently. Until now, smell perception had been

* *Nature*, vol. 449, 27 September 2007

believed to be very much a culturally-driven phenomenon, but this new discovery shows that in fact genes control how we experience certain smells.

So do people who tend to eschew deodorants also tend to carry the altered form of this gene? 'That's something we really want to find out,' says Vosshall. 'In particular we want to know how these different forms of the gene affect human population dynamics.' Which is a polite way of asking whether stinky people marry people with a dodgy sense of smell?

OLDEST HUMANS OUTSIDE AFRICA

Researchers in Tbilisi, Georgia, have uncovered the oldest human remains ever found outside Africa, a species of *Homo* which might even have returned to Africa to spawn modern man.

The Georgian National Museum's David Lordkipanidze and his colleagues*, working at a site in Dmanisi, have uncovered a number of skeletons dating back 1.8 million years. These early people are smaller than modern humans and seem to have features intermediate between the more advanced hominids that gave rise to modern humans, and the earlier *Homo habilis.* Their overall brain and body sizes are small, their hands are more primitive and ape-like, but their legs are more in keeping with advanced species indicating they could travel long distances.

But what's intriguing is that these people clearly pre-date *Homo erectus,* our immediate ancestor, the earliest specimens of which date from about 1.6 million years ago in parts of Ethiopia. So it may be that early hominids like these Dmanisi people, having left Africa many years before, subsequently returned to

* *Nature,* vol. 449, 20 September 2007

what is now Ethiopia to give rise to *Homo erectus*, who in turn evolved into us.

HUMAN BIOLOGY

POLLUTION BLOOD-CLOTTING TRIGGER UNCOVERED

SCIENTISTS HAVE SOLVED a long-running conundrum connecting high levels of air pollution with an increased risk of heart attacks and strokes.

Writing in the *Journal of Clinical Investigation*, Gokhan Mutlu and colleagues* from Northwestern University in Illinois found that mice exposed to airborne particulate matter formed blood clots much more quickly than normal. Tests on the animals' blood showed higher levels of the clotting chemical fibrinogen and elevated coagulation factors II, VIII and X. To find out what was triggering this effect, the team tested mice lacking the gene for an immune-signalling hormone called IL6. Surprisingly, these animals were insensitive to the effects of pollution. Next the team gave normal mice a drug called clodronate to remove a population of cells known as macrophages from their lungs. These cells are phagocytes, meaning that they can ingest foreign inhaled material and when they do so they activate and pump out chemical signals including IL6. The clodronate-treated animals also

* *Journal of Clinical Investigation*, vol. 117, 2007

showed little response to pollution exposure, like their IL6-lacking counterparts.

This suggests that when particulate matter in polluted air enters the lungs it is picked up by macrophages, causing them to activate. The macrophages then pump out IL6, which provokes increases in blood coagulation factors and makes blood much stickier and more likely to clot, which in turn increases the chances of heart attacks and strokes.

NON-STICK CHEWING GUM

CHEMISTS AT THE University of Bristol have come up with a formulation for chewing gum that could make gum-marked pavements a thing of the past. Terry Cosgrove and his team* have developed a new polymer additive that, when added to the gum recipe, renders it water-soluble. 'Chewing gum contains a lot of hydrophobic – water-hating – polymers, which prevent the material from breaking down in water,' explains Cosgrove. This makes it difficult to wash off because the water-repellent quality of the gum prevents water from penetrating and dissolving the material. 'But our formulation contains hydrophilic – water-loving – groups, which, when added to the mixture, allow water molecules to interact with the gum and break it down.'

The result is a material indistinguishable from traditional gum but which disintegrates into fine flakes after prolonged water exposure; but hardcore chewers need not worry, it shouldn't melt in your mouth! Another spin-off is that the new polymer formulation, which is currently undergoing safety tests and is already licensed

* www.bristol.ac.uk/news/

for use in human foods, also intensifies the flavour of the gum by helping the flavourants to interact with water in the mouth, and the taste buds. 'We're hoping to have this on the market within the next year or so,' says Cosgrove.

DRUGS

EXTRA ANTI-CANCER PUNCH FROM TAXOL 'MEATBALL'

US RESEARCHERS HAVE produced a powerful new tool for the delivery of anti-cancer drugs. Using a gold nano-particle core barely bigger than a strand of DNA, Rice University's Eugene Zubarev* has found a way to link dozens of molecules of the drug paclitaxel to the surface of the particle to turn it into a molecular mine.

Normally, administering the right doses of chemotherapy can be problematic owing to difficulties with the solubility of some drugs, and the ability of drugs to penetrate tumours and disperse uniformly throughout the body, and paclitaxel (marketed as Taxol) is no exception. But, because the new nano-balls are loaded with a uniform number of drug molecules, they should help to overcome some of these difficulties. Taxol achieves its anti-cancer effect by binding to structures inside cells called microtubules, which behave like internal scaffolding. To divide, cells need to dismantle and rebuild some of these microtubules, but Taxol stops this from happening by anchoring itself to them using a region at one end of the drug

* *Journal of the American Chemical Society*, vol. 129, 2007

molecule. But this key region can be damaged when attempts are made to anchor the drug to other molecules or nanoparticles. To get around this problem the Rice University team developed a chemical shroud that could protect the Taxol molecules while they were being linked to the gold particle. Once the molecules were in place the shroud was chemically removed.

The next step is to determine the effectiveness of the particles compared with unmodified drug molecules. 'We are already working on follow-up studies to determine the potency of the paclitaxel-loaded nanoparticles,' says Zubarev.

PIN BACK YOUR EARS – BREAKTHROUGH IN AURAL PLASTIC SURGERY

A UK PLASTIC surgeon has come up with a new way to correct ears that stick out too far. Norbert Kang, who's based at the Royal Free Hospital in London, became frustrated with traditional surgical techniques for correcting prominent ears because they are time-consuming, require a general anaesthetic and have up to a 60% complication rate, including scarring and infection. The procedure usually involves stripping the skin from the underlying cartilage and stitching a fold into the back of each ear to keep them 'pinned back'. Instead, Kang has developed a small 20-millimetre by 5-millimetre metal implant, which is inserted between the skin and the ear cartilage through a tiny incision, to bend the ear into the correct shape. It's made from a nickel–titanium alloy, which has a proven safety record because the same material is already used to make angioplasty stents for propping open blocked arteries. This metal also has 'shape memory', meaning that it can easily be bent or deformed but springs back to its original position when released, which prevents the ears from being deformed by knocks and scrapes. According to Kang, the implant can be inserted in under ten minutes using just local anaesthetic, which makes the procedure significantly cheaper and safer than existing practices, although it still needs evaluating to prove that it has a satisfactory success rate.

CHEMISTRY

COOL IT

US ENGINEERS HAVE developed a system to keep houses cool without turning up the aircon. Bill Miller and his team* from Oak Ridge National Laboratory in Tennessee have come up with a roofing system that soaks up heat during the day and then re-radiates it out to space at night, keeping homes cool in the process.

There are four parts to the new system. The first three are concerned with reflecting heat off the roof surface using a combination of reflective tile surfaces, modified tiles that channel warm air out from within the roof cavity, and further reflective coatings beneath the rafters. These are fairly standard techniques, but the fourth line of defence is highly innovative. It consists of sheets of a material which melts at temperatures above 23° Celsius. Melting a solid provokes a phase change, which consumes energy, so the material remains at the same temperature until it has all melted, which can take all day. In this way the material behaves as an energy sponge, preventing the heat from entering the living space below. Later, when the temperature falls at night, the material reverts to a solid again, releasing

* www.ornl.gov/info

the energy it soaked up during the day and radiating it out into space.

During tests, the material reduced attic temperatures on sunny days by more than 5° Celsius. 'We're able to intercept 90% of the heat energy that would otherwise penetrate into the living space through the attic floor. This could reduce the cooling bill for houses by up to 8%,' says Miller.

HUMAN BIOLOGY

ONLY THE LONELY – GENOMICALLY SPEAKING

SCIENTISTS IN AMERICA may have found out why loneliness is linked to health problems. Writing in the journal *Genome Biology*, Steve Cole and his colleagues* report how they collected blood samples from 14 volunteers who had been matched for age, health status, weight, and medication use. The only difference was that six of the subjects scored in the top 15% on a UCLA Loneliness Scale. The rest of the group were in the bottom 15% on the same scale. The team analysed the blood samples to study patterns of gene expression from the 14 subjects and found 209 genes that were consistently differently expressed between the lonely and the contented subjects. Of those 209 genes, 78 were genes that were more active, and 131 were genes that were less active. 'White blood cell gene expression appears to be remodelled in chronically lonely individuals,' said Cole.

This finding might explain why people who feel socially isolated tend to have a higher risk of certain diseases, such as heart disease, high blood pressure, and susceptibility to infections and cancer, compared

* *Genome Biology*, vol. 8, 2007

with individuals with a stronger social network. 'These findings provide molecular targets for our efforts to block the adverse health effects of social isolation,' says Cole. But the health-promoting effect of friends isn't just down to how many people you know. 'We found that what counts at the level of gene expression is not how many people you know, it's how many you feel really close to over time.'

In the future, he suggests, doctors might be able to use the team's genetic fingerprint of loneliness to monitor the effect of interventions intended to reduce the impact of a poor social network on health.

GEOLOGY

THE ORIGIN OF ARMAGEDDON ASTEROID UNCOVERED

SCIENTISTS IN THE US and the Czech Republic have discovered the origins of the meteorite that helped bring about the demise of the dinosaurs.

Bill Bottke from the Southwest Research Institute in Boulder Colorado, together with David Vokrouhlicky and David Nesvorny* from Charles University in Prague, used numerical simulations to wind back the cosmic clock over 160 million years to piece back together the puzzle of what plunged the planet into the equivalent of a nuclear winter and wiped out the dinosaurs. The team used computer simulations to track the orbits of these fragments back to the time when they first formed. Their calculations show that a large 170 kilometre-diameter asteroid, known as the Baptistina parent body, collided with a smaller 70 kilometre object between Mars and Jupiter about 160 million years ago. The result was the creation of more than 1000 smaller asteroid pieces each measuring 1 kilometre or more in diameter.

This occurred in a part of the solar system where the combined gravitational effects of Mars and Jupiter

* *Nature*, vol. 449, 6 September 2007

create what Bill Bottke terms a 'celestial escape hatch'. Here, warming by the sun's rays triggers a process called the YORP effect, which can alter the path of small objects and nudge them into Earth-crossing orbits. The result was a 100-million-year-long meteor shower, which, say the researchers, included the 10 kilometre-diameter 'Chicxulub' impactor that struck Mexico's Yucatan Peninsula about 65 million years ago. And a recently discovered meteorite fragment recovered from the Pacific in sediments dating back to the same era is a chemical match for the Baptistina fragments still in orbit out beyond Mars.

'So we can say with more than 90% probability that this breakup event 160 million years ago looks like the origin of the impactor that produced the mass extinction event 65 million years ago,' says Bottke.

HUMAN BIOLOGY

START 'EM YOUNG – 'CIGARETTE WITH YOUR MORNING BREAST MILK?'

US RESEARCHERS HAVE found altered sleep patterns amongst breastfed infants of mothers who smoke. Julie Mennella* from the Monell Chemical Senses Centre in Philadelphia recruited 15 breastfeeding mothers who smoked. Sleep and activity patterns in their babies, which were aged between two and six months, were monitored on two occasions over a three-hour period after the children were fed. On both occasions the mothers were asked to abstain from smoking for 12 hours before the study, but on one of the two occasions they were allowed to smoke just before they fed their babies. The women were also asked to avoid caffeinated drinks during the study.

Tests on the milk from mothers who had recently smoked confirmed that the babies were receiving a significant increase in nicotine dose, and the team found that the amount of sleep taken during the following three hours by these babies fell from an average of 85 minutes to 53 minutes, a drop of almost 40%. This is probably due to the neuro-stimulatory effects of nicotine, which has been shown to inhibit

* *Pediatrics*, vol. 120, no. 3, September 2007

regions of the brain which are concerned with controlling sleep. It may also, suggests Manella, explain why neonatal nicotine exposure has been linked in the past with long-term behavioural and learning deficits, since these could be the consequence of sleep disturbance.

In light of these findings, mothers who smoke might want to consider planning their smoking around their breastfeeding. Nicotine levels in milk peak 30–60 minutes after smoking, but take three hours to return to baseline, so this might be feasible.

NON-HUMAN BIOLOGY

JAWS II – THE MORAY RETURNS

SCIENTISTS HAVE DISCOVERED how moray eels manage to swallow things nearly as large as themselves – they have a second set of teeth in their throat, called pharyngeal jaws, which jump forwards and grab whatever's in the mouth and drag it down the animal's oesophagus.

Rita Mehta*, from UC Davis, made the discovery by capturing fast camera footage of morays as they dined, and then confirmed the findings with fluoroscopy, the eel-equivalent of a barium swallow. 'These creatures have evolved this intriguing adaptation because their large body size makes them dependent on eating other large creatures like fish and cephalopods,' explains Mehta. 'Most fish use suction to pull prey into their mouths and down their throats; but larger prey can more easily escape, and since morays live in crevices and holes, they would not be able to expand their heads to create the suction they needed to capture things.'

Other long thin creatures, like snakes, use alternative methods to swallow large prey, such as dislocating their jaw and ratcheting it left to right to squeeze

* *Nature*, vol. 449, 6 September 2007

things in. But this is the first time that an alternative to hydraulics has been seen in fish.

DISEASES AND DISORDERS

DIABETES ALL IN THE MIND

WELL NOT QUITE all, but new research suggests that a population of brain cells have a role to play in triggering the non-insulin-dependent form of the disease. Writing in *Nature*, Laura Parton and her colleagues* at Harvard describe how they have identified glucose-sensitive neurones in the brain's hypothalamus which become excited when blood sugar levels rise. Since the hypothalamus controls energy balance and metabolism, the researchers wondered whether these glucose-sensing cells could also play a role in diabetes.

To find out, they genetically deactivated the glucose sensor used by these nerve cells and studied how this affected blood chemistry in a group of experimental mice. Despite each having a perfectly normal pancreas capable of making normal levels of insulin, the animals developed a condition similar to the human pre-diabetic state known as 'impaired glucose tolerance'; in other words, they couldn't control their blood sugar levels correctly. Next the team explored what role these cells might play in the association between diabetes and obesity. Two groups of identical, normal

---

* *Nature*, vol. 449, 13 September 2007

mice were given either a healthy diet or fed the rodent equivalent of high-calorie junk food.

As expected, the overfed mice became obese, and when they did so the glucose-sensing cells in their brains stopped responding normally to blood sugar. The mice of the same age on the healthy diet, meanwhile, showed normal responses. Since shutting off these cells produced a diabetic-like state in the first group of mice, it may be that this is what links obesity with diabetes. 'The discovery of defects in glucose-sensing by the brain could help lead to new therapeutic strategies,' points out Brad Lowell, who led the study.

ECOLOGY

PLANTS LESS THIRSTY IN A WARMER WORLD

UK RESEARCHERS HAVE found that rising carbon dioxide (CO_2) levels could have potential benefits for drier parts of the world – because plants become less thirsty. Richard Betts and his colleagues* from the Hadley Centre in Exeter have found that a doubling of CO_2 concentrations compared with pre-industrial levels, which is predicted within the next 50–100 years, will result in 6% more water in rivers because of more runoff from the land. This greater availability of water occurs because higher CO_2 levels cause plants to use less water. The undersides of a plant's leaves are peppered with tiny pores called stomata, which can be opened to allow plants to soak up the CO_2 they need for growth from the atmosphere. The CO_2 enters the leaf by dissolving in a layer of water spread over the surfaces inside each of the stomata. But this water constantly evaporates into the air in a process called transpiration, which is why plants need a continuous supply from their roots.

Higher CO_2 levels, however, mean that plants don't need to open their stomata as wide or for as long to

* *Nature*, vol. 448, 2007

obtain the same amount of carbon dioxide, which cuts down their water usage. And what the plants don't drink then ends up back in rivers.

The team reached their conclusions, which are published in *Nature*, by using the same climate models which are used to predict day-to-day weather forecasts and which are therefore well tested and viewed as reasonably accurate. 'This indicates that freshwater resources may be less limited than previously assumed,' says Betts.

ECOLOGY

PPPP-PICK UP A POO-LUTANT

Belgian scientists have found that penguin poo is causing concentrations of pollutants to build up in the Antarctic. Adrian Covaci and colleagues* from the University of Antwerp collected samples from around the nesting sites of Antarctic non-migratory Adelie penguins and found unexpectedly high levels of organic pollutants. These chemicals, which were at concentrations 10–100 times greater than anticipated, come from man-made sources including brominated flame retardants and organochlorine pesticides. They are normally carried in the air and by ocean currents, so why should they have reached such high levels around the penguin colonies? The answer may be bio-accumulation – the concentration of chemicals by the food chain. The penguins are catching and eating fish, which pick up traces of the pollutants from the seawater. But by eating large numbers of fish the penguins increase their intake considerably. Back ashore, their guano and the carcasses of their dead, which accumulate around their nesting sites lead to a build-up of the chemicals in the soil. Commenting on the study,

* *Journal of Environmental Monitoring*, vol. 9, 2007

UK organic pollutant researcher Kevin Jones from the University of Lancaster says these results are important because they highlight an unusual mechanism that is moving chemicals around the planet, particularly since the long-term consequences of a build-up like this are not known.

CHEMISTRY

WORLD'S SMALLEST THERMOMETER – A SINGLE MOLECULE

RESEARCHERS HAVE FOUND that a green-glowing chemical from a jellyfish can be used as a molecular thermometer. Known as GFP (green fluorescent protein), the substance is used by biologists to label cells harmlessly. Shining light of a certain wavelength at a cell containing GFP causes the molecule to flash up green. But a closer look at the green light given out reveals that it is not continuous and instead rapidly blinks on and off as the molecule subtly alters its structure. Now Cécile Fradin and her colleagues* from McMaster University in Hamilton, Ontario, Canada, have found that the rate of blinking corresponds to the temperature: when things get hotter the blinking slows down, at cooler temperatures the blinking speeds up.

To find out whether this could be used as a miniature optical thermometer the team measured the blink rate at different temperatures between 10° and 50° Celsius. They report in the *Journal of the American Chemical Society* that they found with this technique they could make measurements to within an accuracy of 1°. Previously, researchers had tried to use dyes which glow

---

* *Journal of the American Chemical Society*, vol.129, no. 34, 2007

more brightly at different temperatures, but these can be misleading if several molecules congregate together producing a brighter light. The new discovery is likely to prove useful for temperature-monitoring applications on miniature lab-on-a-chip diagnostic devices, and even for measuring the temperature of different structures inside living cells.

NON-HUMAN BIOLOGY

PLANTS CAN HEAR – SO IT MIGHT MAKE SENSE TO TALK TO THEM AFTER ALL

KOREAN RESEARCHERS CLAIM to have discovered two sound-sensitive genes in rice plants. Mi-Jeong Jeong and colleagues* from the National Institute of Agricultural Biotechnology in Suwon, South Korea, made the discovery by exposing plants to noise and studying how this affected gene activity.

Initially, they tried playing 14 samples of classical music to their rice, including Beethoven's *Moonlight Sonata*, but saw no response. Then they tried playing sounds of specific frequencies and began to see a change. At 125 Hz and 250 Hz the activity of two genes, *rbcS* and *Ald*, went up. When sounds at 50 Hz were played the gene activity went down. Then, to find out whether other genes could be rendered sound-sensitive, the team linked the genetic elements that control the Ald gene to a 'reporter' gene inserted into the rice called GUS (beta-glucuronidase). They found that sounds could then also manipulate the levels of GUS. This suggests that it might be possible to use sound instead of chemicals to control different genetic processes in plants, such as switching on

* *Molecular Breeding*, vol. 21, no. 2, July 2007

resistance genes to help them fight off pests or ripen more rapidly.

Not all scientists are convinced, though; some say that the new research, which is published in the journal *Molecular Breeding*, makes use of old techniques and lacks sufficient numbers to make the data credible. Sounds like sour grapes, although it does suggest that it's not just cereals like wheat and barley that have 'ears'.

MAKE NO BONES ABOUT IT: THE THEORY OF EARLY MAN WAS WRONG

THE DISCOVERY OF two new fossils looks set to rewrite the history of human origins. Existing theories suggest that the early hominid *Homo habilis*, which first appeared about 3 million years ago, slowly morphed into the bigger-brained *Homo erectus*, which then turned into us, *Homo sapiens*. But now, working in north-east Kenya, University College London's Fred Spoor and his colleagues* have uncovered two 1.5 million-year-old fossilised skulls that rubbish that theory and also provide us with new insights into the lives of our ancient ancestors.

Morphological features prove that one of the new fossils is from a *Homo habilis*, the other from a *Homo erectus*. The team used the age of the sediments in which the fossils were found to date the specimens and were surprised to see that the more 'modern' *Homo erectus* remains, at 1.55 million years, were older than those of the *Homo habilis*, which clocked in at 1.44 million years. These dates mean that, contrary to the theory that one species spawned the next, both *Homo habilis* and *Homo erectus* must have been co-existing

* *Nature*, vol. 448, 9 August 2007

side by side in this same part of Africa for over half a million years.

'Their co-existence makes it unlikely that *Homo erectus* evolved from *Homo habilis,*' says Meave Leakey, one of the study co-authors. Instead it looks like both evolved independently between 2 and 3 million years ago, although from what remains an as-yet-unsolved puzzle.

FOOD

DIET FOODS MAKE CHILDREN FAT

US RESEARCHERS HAVE found that giving young animals diet foods can trigger obesity by encouraging overeating behaviours, suggesting that the same thing could happen to young children.

Writing in the journal *Obesity*, University of Alberta researcher David Pierce and his colleagues* have found that giving young rats the rodent equivalent of low-calorie foods affected their ability to learn how to associate the amount of energy in food with its taste. It seems that artificial sweeteners can fool the brain into assuming that all sweet things are equivalently low in calories, which can trigger overeating when other non-diet foods are consumed. In the present experiments, the age of the animals also seemed to be critical, because adolescent rats were not affected, possibly because they had already established the relationship between tastes and calories.

'One thing is clear at this point,' says Pierce. 'Our research has shown that young animals can be made to overeat when low-calorie foods and drinks are given to them on a daily basis, and this subverts their

---

* *Obesity*, vol. 15, 2007

bodies' energy-balance systems.'

The results also fit with other recent findings including a University of Massachusetts study which showed links between diet drink consumption (amongst children) and obesity, diabetes and heart disease.

'Diet foods are probably not a good idea for growing youngsters,' Pierce cautions.

PHYSICS

NUCLEAR SPACE SCREW TACKLES NEOS

ALTHOUGH SPACE SCIENTISTS are confident that they have mapped the majority of 'near Earth objects' (NEOs) that could conceivably collide with us, and found that we're safe for now, there's always the possibility that something unexpected might happen. Like the asteroid Apophis, which will slip past the Earth in 2029 and then make a return visit in 2036. If it alters its course, there is a remote one in 45,000 chance that it could hit us.

Thankfully, University of Rome La Sapienza researcher Daniele Fargion has come up with a new strategy to tackle the problem head on: a nuclear-powered rocket drill that burrows into the object and hurls the rock it digs out into space, pushing the asteroid off its Earth-bound course. This is an application of Newton's Third Law – for every action there is an equal and opposite reaction. Flinging the drilled-out debris into space will give the object a kick in the other direction. Fargion's calculations show that over a 10-year period this could deflect a 1 kilometre-wide asteroid by up to 30,000 kilometres, enough to miss the Earth.

But it's not all plain sailing. Asteroids are often loose aggregations of rocks and debris, which may prove difficult to drill into. Another problem is that they are also very often spinning, so the system would have to be programmed to spit out rocks only when the asteroid was pointing in the right direction. Undeterred, however, Fargion proposes testing the idea on our own moon, where his 'screw rockets' could help to dig out underground shelters for future human use.

METEOROLOGY

SPOT THE RAIN

RESEARCHERS HAVE SPOTTED a pattern in the floods that hit parts of Africa – they tend to occur in the year preceding a peak in sunspot activity. Writing in the *Journal of Geophysical Research*, Curt Stager from Paul Smith's College, New York, and his colleagues* looked at 100 years of rainfall data and water-level records from Lakes Victoria, Tanganyika and Naivasha, which showed that rainfall peaks occurred in the years immediately before the 11-year sunspot peaks. The next peak is expected in 2011–2012, meaning that if the relationship holds true, as it has done throughout the 20th century, then rainfall should peak again in 2010.

The researchers suspect that the increased solar energy associated with sunspots leads to increased heating of both the land and sea, which promotes moisture evaporation and precipitation. It could also be linked to El Niño, which increases rainfall in east Africa. The finding may help authorities to plan their resource allocations better to tackle the problems commonly associated with heavy rainfall, chiefly

* *Journal of Geophysical Research*, vol. 112

flooding, erosion and disease, particularly water-borne illnesses and malaria.

'When you think of climate troubles in Africa, it's usually about drought,' says Stager. 'You don't often think of the opposite situation. Too much rain can create just as many problems.'

ECOLOGY

LIFE UNDER A BROWN CLOUD – BUT IT DOES HAVE A SILVER LINING

SCIENTISTS HAVE DISCOVERED that the global warming gas CO_2 has an accomplice – clouds of soot in the atmosphere known as brown clouds. These clouds, which are over 3 kilometres thick, are produced by industry, traffic and fires and contain a mixture of carbon particles and oxides of nitrogen and sulphur.

Scientists had thought that these clouds mainly cool the planet by reflecting sunlight back into space, perhaps mitigating the effects of the greenhouse effect by up to 50%, but their effect on the local environment wasn't known. To find out, Vee Ramanathan*, from the University of California at San Diego, sandwiched one of these clouds in southern India between three unmanned aircraft, which were equipped with instruments to measure temperature and light intensity. The readings showed that the soot particles were capturing heat from the sun and transferring it to the atmosphere, warming things up locally.

'We found that the brown cloud enhanced solar heating by around 50%,' says Ramanathan. And the effect is considerable: 'The warming from brown

* *Nature*, vol. 448, 2 August 2007

clouds is about the same as the warming from recent rises in greenhouse gases.' This means that the heating effect may be sufficient to explain the retreat of Himalayan glaciers in recent years.

But even brown clouds have a silver lining, it seems, because they last for only 2 weeks before dissipating. So, if steps are taken to prevent their formation, it might be possible to reduce the rate of Himalayan melting.

DRUGS

A RASH OF NEW ANTI-HIV DRUGS

A CLUSTER OF new anti-AIDS drugs are expected to be approved for general use, giving hope to patients who have developed drug-resistant forms of the virus*. This is because the new agents hit different parts of the virus compared with existing anti-HIV drugs, which means that there should be very little initial drug-resistance.

The new agents include two 'fusion inhibitors', maraviroc, made by Pfizer, and vicrivoroc, which has been developed by Schering Plough. These drugs block the ability of HIV to lock onto a molecule on the cell surface called CCR5, which the virus needs in order to penetrate and infect.

The other agent, raltegravir, is made by Merck. It locks onto a viral protein called integrase and prevents it from inserting a copy of its genetic material into the host DNA, which is a critical step in the viral lifecycle. When both of these approaches have been combined (a fusion inhibitor given together with the integrase inhibitor) in patients with drug-resistant disease, the

---

* *European Journal of Medical Research*, vol. 12, 15 October 2007
Expert Review of Anti-infective Therapy, vol. 6, no. 4, August 2008

viral load in the bloodstream fell to undetectable levels, indicating good disease control.

SATNAV RECEIVES ENVIRONMENTAL THUMBS UP

WHAT'S BEST FOR the environment, a well-thumbed map and some common sense, or a satnav? To find out, Taiwanese researchers Wen-Chen Lee and Bar-Wen Cheng* from the National Yunlin University of Science and Technology recruited 32 drivers and asked them to navigate to a series of predetermined locations; half of them were asked to get there by satnav, the other half had to use a map. The team monitored the progress of the drivers and clocked the distances they travelled.

In all cases the satnav triumphed. On urban routes, journeys were on average 7% shorter than when the driver followed a map, and even in the open country routes remained 2% shorter. Since shorter trips on average use less fuel, they are more environmentally friendly.

But satnav may also improve driving safety, Lee points out, because the map users consistently changed course more times per journey than their satnav-guided colleagues. More course changes can be a sign of frustration, which can lead to dangerous driving.

---

* *Accident Analysis & Prevention*, vol. 40, Issue 1, January 2008

FRENCH BIKE THEIR WAY TO CLEANER AIR

PARIS IS ONE of Europe's most congested cities, but now officials may have a new weapon to combat the problem – free bikes. The city has set up an initiative called 'Velib' and wheeled out 10,000 distinctive grey-green bikes which registered users can pick up and take for a spin.

To access the bikes, riders purchase an access card – one day costs 1 euro, a weekly card is 5 euros or an annual card is 29 euros. Armed with a card, riders can enjoy the first half-hour for free, and then a rising supplement is charged for each additional half-hour. The idea is that most people should be able to get where they need to go within the free half-hour and the rising cost penalty should keep lots of bikes in circulation. There are already 600,000 registered riders and 750 bike pick-up and drop-off points along Paris's 370 kilometres of cycle paths; by the end of the year the city aims to have 20,000 bikes in service and nearly 1500 pick-up points. But not everyone's enamoured with the new initiative. Taxi drivers, perhaps not surprisingly, are grumbling about having to avoid cyclists enthusiastically pedalling the wrong way up one-way streets, and pedestrians are having to dodge bikes that have invaded the pavements.

And human nature has also created something of a problem with supply and demand – lazy Parisians are picking up bikes at the tops of hills, coasting to their destination and then taking the Metro back!

HUMAN BIOLOGY

RESEARCHERS SCRATCH THE SURFACE OF UNDERSTANDING ITCHING

RESEARCHERS HAVE UNCOVERED a gene that transmits the itch sensation. The result means that drugs capable of providing the pharmacological equvalent of a 'scratch' could soon be on the way, sparing pruritic patients the misery of chronic itch disorders like eczema. Writing in *Nature*, Yan-Gang Sun and Zhou-Feng Chen* from Washington University in St Louis describe how they homed in on a small population of sensory nerves that release a transmitter substance called GRP – gastrin-releasing peptide. GRP locks onto a chemical docking station in the spinal cord called GRPR – the gastrin releasing peptide receptor. The researchers found that when they 'knocked out' the gene coding for GRPR in mice, the animals became much less susceptible to itch-provoking stimuli than normal mice, but were otherwise normal. The team also found that normal mice could be made itch-resistant by injecting a chemical that blocks GRPR into the spinal fluid, confirming its role in the itch-sensing pathway. However, it's almost certain that there are some other itch pathways still waiting to be uncovered.

---

* *Nature*, vol. 448, 9 August 2007

'The fact that the knockout mice still scratched a little suggests there are additional itch receptors,' Chen points out.

But the good news is that GRP has been studied previously in connection with certain types of cancer, so there are already a number of drugs available that are known to block it. 'So now researchers can study the effect of these agents on the itch sensation and possibly move that research to clinical applications fairly soon,' says Chen.

HUMAN BIOLOGY

DIESEL EXHAUST AT THE HEART OF ARTERIAL DISEASE

RESEARCHERS HAVE FOUND that airborne pollution can trigger damage to blood vessels. Ke Wei Gong and colleagues*, from the University of California at Los Angeles, cultured endothelial cells of the type that line blood vessels with particles from diesel exhaust and oxidised phospholipids of the kind associated with LDL or 'bad' cholesterol. The cells were then analysed to study the patterns of genes that had been switched on or off in response to the exposure, compared with unexposed controls.

The researchers found the diesel exhaust particles were affecting at least three genetic pathways linked to inflammatory processes in the linings of blood vessels. Next they exposed mice, which had been genetically programmed to develop the rodent equivalent of high cholesterol levels, to the diesel particles. They found the same pattern of altered gene activity in the animals as they had seen in the cell cultures.

These results may explain the observed link between heart attacks and strokes and levels of atmospheric

---

* *Genome Biology*, vol. 8, 2007

pollution, and also highlight a mechanism by which pollution might increase the risk of vascular diseases.

MEDICAL TECHNOLOGY

LAB ON A CHIP FOR IVF

JAPANESE RESEARCHERS ARE developing a 'lab-on-a-chip' to improve the success rates of IVF techniques. Researchers think that manipulating eggs and sperm, and the un-physiological (abnormal) environment of the dish in which fertilisation and initial development occurs, may have an adverse effect on the success of IVF.

To combat the problem, Teruo Fujii,* from the University of Tokyo, has produced a microfluidic device which can take 20 eggs, mix them with sperm and then nurture them as they are fertilised and begin to develop. To create a more ideal environment the team grow a carpet of endometrial cells – which normally line the uterus – in their device. This means that growth factors and other signals from these cells can reach the developing embryo, more closely mimicking the environment in the body. Initial tests using animals have been encouraging. Out of 50 eggs fertilised on the chip, 30 successfully developed into early embryos compared with 26 out of 50 produced the traditional way. In a further study, when the embryos

* www.newscientist.com

were placed in the uterus to test their long-term viability, 44% of those produced on the chip grew into mice, compared with 40% of those produced using existing IVF approaches. Although the improvement appears quite modest, in a high-stakes game like human infertility, 4% can make the difference between joy and heartbreak.

The team now have permission to test their approach using human embryos, which they hope to start doing later this year.

CHEMISTRY

FRIEND, OR 'FOE-ZONE'

RESEARCHERS HAVE FOUND that ozone, the chemical we can't live without, could also be the death of us, or at least a potent trigger for global warming.

Stephen Sitch*, from the UK's Hadley Centre, has found that ozone at ground level interferes with the ability of plants to soak up carbon dioxide, meaning future CO_2 predictions could have been significantly underestimated. Ground-level ozone, which is produced when sunlight interacts with airborne pollutants, has been steadily increasing in concentration in recent years. It enters the leaves of plants via their stomata – mouths in the undersides of the leaves that allow CO_2 from the atmosphere to enter. Once inside the plant, it damages cells and stunts growth, reducing the amount of carbon dioxide the plant subsequently removes from the air.

'Although increased CO_2 in the atmosphere will have a "fertilising" effect on plants, rising ozone will nevertheless cut the amount of CO_2 removed from the atmosphere. So we may have underestimated the size of the warming effect we're going to get,' says Sitch.

---

* *Nature*, vol. 448, 26 July 2007

GEOLOGY

FLOOD DISCOVERY MAKE WAVES OVER ORIGINS OF ENGLISH CHANNEL

RESEARCHERS BASED IN London have uncovered the event to which Britain owes its island existence. Writing in *Nature*, Sanjeev Gupta and Jenny Collier*, who are both based at Imperial College, used sonar to scan the floor of the English Channel looking for clues to what cut the UK adrift from France.

What they uncovered was evidence of a catastrophic flood that carved out the Straits of Dover at some point between between 200,000 and 400,000 years ago. At this time England was linked to the continent by a rock ridge made of chalk that ran between Dover and France. This meant that the Thames and the Rhine both flowed northwards to empty into the North Sea.

Ice sheets triggered by the arrival of a glacial period formed an ice dam, which blocked the flow of water and triggered the formation of a lake the size of Wales. The water backed up all the way to the rock ridge. The team can't say for sure what triggered it, and it could have been a small earthquake, but something suddenly caused a breach in the ridge, which quite

* *Nature*, vol. 448, 19 July 2007

literally unleashed the floodgates. The accumulated water surged through, carving out deep chaotic gouges in the chalk, which were the giveaway signs picked up by the team's sonar.

'What's really exciting about this result is that it ties in with what the archaeologists are finding about early humans in Britain,' says Sanjeev Gupta. 'Between 200,000 and 60,000 years ago there is no evidence of human activity here; so it could be that because this flood triggered the Rhine to divert south through the newly formed Straits of Dover it would have been very difficult to cross, and as sea levels rose with the end of the glacial period [ice age] Britain would have been cut off.'

NON-HUMAN BIOLOGY

GECKOS 'MUSSEL IN' ON UNDERWATER ADHESIVE SCENE

RESEARCHERS HAVE SUCCESSFULLY combined two of nature's most powerful adhesive strategies to produce the underwater equivalent of a Post-it™ note. Writing in *Nature*, Phil Messersmith* from Northwestern University in Illinois describes how he and his team have come up with 'Gekel', which consists of an array of silicone nanorods designed to mimic a gecko's foot, coated in a polymer recipe based on the adhesive used by mussels to glue themselves onto rocks and jetties.

Geckos are the lizard equivalent of spidermen. They can run up vertical surfaces and even move across surfaces upside down. The secret to their surface-hugging success is an array of microscopic nano-hairs on their feet. These tiny hairs, each no more than 1 five-thousandth of a millimetre across, are electrically attracted to the surface they are touching, giving the gecko sticking power. But unfortunately geckos come unstuck in the wet, which has thrown a spanner in the biomimetic works of researchers who are trying to copy the gecko's trick for use in robotics and to produce better adhesives.

---

* *Nature*, vol. 448, 19 July 2007

This problem led Messersmith to wonder whether mussels might be the answer. These shellfish produce proteins containing large amounts of the unusual amino acid DOPA (di-hydroxy-phenylalanine), which sticks tightly to most surfaces. So he produced an artificial version of the mussel glue and smeared it onto the artificial gecko foot. The result was an adhesive that can be attached, detached and reattached at least 1000 times without losing its stickiness, in and out of the water.

Now the team are looking at ways to scale up production of the new material, and they hope one immediate application will be in producing better wound dressings, and plasters that won't float off in the pool.

SWEET TASTE OF SUCCESS AS RESEARCHERS UNCOVER GENE THAT CAUSES DIABETES

RESEARCHERS HAVE FOUND a new gene that is strongly linked to the development of type 1 or 'juvenile' diabetes.

Hakon Hakonarson*, from the Children's Hospital of Philadelphia, used the power of SNPs – single nucleotide polymorphisms – to track down the gene in children with the disease. SNPs are genetic markers that are inherited like any other piece of DNA, but because they have all been documented they can be used as convenient flags to highlight the presence of certain genetic sequences. In this study the researchers recruited 1000 diabetes patients, 1200 parents of patients with diabetes and 1000 healthy children and studied the pattern of SNPs in all of them.

This led to them identifying all of the previously discovered genetic triggers of diabetes as well as one new one, a gene called KIAA0350. It's carried in at least 40% of children with diabetes and although the team don't yet know exactly how it triggers the

* *Nature*, vol. 448, 2 August 2007

disease, it is strongly expressed by immune cells. This is significant because type 1 diabetes occurs when the immune system turns upon and destroys a patient's own insulin-producing cells in their pancreas.

But now the team have identified the gene, which was previously unknown, they can begin to ask important questions about how it provokes the condition, and whether it can be deactivated to protect carriers from subsequently developing the disease.

ANTHROPOLOGY

MODERN HUMAN'S 'GARDEN OF EDEN' CONFIRMED AS AFRICA

WHERE WE MODERN humans all came from has been a subject of intense debate. Some parties suggest that modern humans sprang up in several places around the globe, while others have steadfastly insisted that, like Karen Blixen, we all came out of Africa.

Now scientists have produced the strongest evidence yet that this latter argument is the correct one and that southeast Africa was the cradle out of which modern mankind climbed. Cambridge University's Andrea Manica and his colleagues* found the genetic diversity of different populations around the world declined the further those populations were from Africa. They then measured the variation in the sizes and shapes of over 4000 human skulls representing 105 populations worldwide from the last 2000 years, and found an identical relationship. The greatest variation in skull structure was found in samples from east Africa.

The loss of diversity with increasing distance from Africa is the consequence of population 'bottlenecks' – abrupt reductions in population size during migration due to adverse conditions. The further a

* *Nature*, vol. 448, 2 August 2007

population migrated the greater the number of these bottlenecks a population would have faced and hence the lower their diversity.

To prove that they were on the right track the team tried to fit their data to other sites around the world that could have spawned modern man. 'This just did not work. Our findings show that humans originated in a single area in sub-Saharan Africa,' says Cambridge co-author Francois Balloux.

HUMAN BIOLOGY

HAVE A BREAK. HAVE A KITKAT . . . WELL MAYBE NOT, BUT YOU WILL LOSE MORE WEIGHT

JAPANESE RESEARCHERS HAVE found that when it comes to exercise and weight loss, the perceived wisdom of quantity over quality might be wrong. Tokyo University's Kuzushige Goto* recruited six healthy men and studied their metabolic response to exercise on a cycling machine. The volunteers performed either a solid 60-minute workout, two 30-minute workouts with a 20-minute rest between the two, or 60 minutes resting in an armchair as a control. Blood samples were taken regularly throughout the experiments.

Surprisingly, the men showed signs of burning off more fat during the interrupted exercise regime than during the sustained 60-minute workout. Their blood showed significantly higher levels of free fatty acids and glycerol, which are markers of fat breakdown, compared with the longer workout.

Present advice given to patients trying to lose weight by exercising is to extend the length of workouts to ensure adequate fat depletion. However, this study

---

* *Journal of Applied Physiology*, vol. 102, 2007

shows that this method may not be the most effective way to enhance fat metabolism, as splitting up a long bout of exercise with a rest period burns more fat than a continuous bout of exercise.

ASTRONOMY

RESEARCHERS CONFIRM THE EXISTENCE OF WATER ON AN EXTRASOLAR PLANET

FOR THE FIRST time researchers have been able to say with certainty that there is water on a distant planet. Writing in *Nature*, University College London's Giovanna Tinetti and her colleagues* used NASA's Spitzer Space Telescope to study a 'hot Jupiter' orbiting a star 60 light years away. Like our own Jupiter, the planet is a gas giant but, unlike our solar system, it sits very close to the parent star, completing an orbit in just 2.2 days.

The team were able to watch as the planet eclipsed the star on each orbit and look for the infrared fingerprint of water in the starlight passing through the planet's outer atmosphere. 'We can now say with some security that the water signal is definitely there,' says Giovanna Tinetti. But it's unlikely that the planet will become a holiday destination any time soon. It's so close to the parent star that the sunlit side is a sizzling 1200° Celsius and even the dark side is a roasting 800° C. As a result it's too inhospitable for life but, as Tinetti points out, 'Our discovery shows that water might be more common out there than previously

* *Nature*, vol. 448, 12 July 2007

thought. Our method can be used in the future to study more 'life-friendly' environments.' The same planet also made headlines when scientists were able to forecast its weather.

ELECTRICITY

iLIGHTNING

A PAPER IN the *New England Journal** describes a man admitted to hospital with a rather strange pattern of skin injuries including ruptured eardrums, a broken jaw and burns to his chest, neck and the inside of both ears!

Doctors discovered that the 37-year-old had been out jogging in a thunderstorm, while listening to his iPod. Lightning had hit a nearby tree and jumped onto the man as he ran past, a phenomenon known as side flash. This triggered muscle contractions that threw him 2.4 metres. The man's burns ran in two lines up his neck, across the sides of his face and entered his ears. They corresponded to the position of his headphones when the accident happened.

Although Eric Heffernan and his colleagues, who described the case, are at pains to emphasise that iPods aren't a specific risk factor for being hit by lightning, in this instance the combination of sweat, metal wires and headphones channelled the electricity into the patient's head. The sudden heating effect caused by the discharge into the ears caused rapid expansion of

* *New England Journal of Medicine*, vol. 357

the air in the auditory canal, bursting the patient's eardrums. It's not known what he was listening to on the iPod at the time the bolt hit, but my money's on Pink Floyd's 'Delicate Sound of Thunder'.

FLY TOOLBOX

A POWERFUL NEW tool for geneticists was announced in *Nature* in the form of a comprehensive genetic library that can be used to shut off any gene in any tissue of a fruit fly. It works by using a technique called inducible RNAi (RNA interference) to shut off specific genes in specific cells and tissues at specific points during development.

This means that, for the first time, researchers will be able to tease out what individual genes do in individual cells without having to 'knockout' the gene from the entire animal (which causes all kinds of confusing side effects). Barry Dickson* from the Research Institute of Molecular Pathology (IMP) in Vienna says his work will inevitably lead to considerable breakthroughs, given the precision with which researchers will now be able to study the action of specific genes.

The work also has direct relevance to humans, since there is a significant genetic overlap between many of the systems of flies and men, meaning that flies can

* *Nature*, vol. 448, 12 July 2007

provide clues to the workings of some human genes and tissues.

ECOLOGY

BACTERIA ON HUNGER STRIKE

A PAPER IN *Nature** contained worrying news for waterways. Bacterial communities living in estuaries usually protect the sea from pollution by removing the excess nitrogen washed in from farmland. But since the 1970s the amount of nitrogen removed has been falling. Now it's reversed and the riverbed bacterial community seems to be adding nitrogen to the water rather than taking it away.

Robinson Fulweiler, from Rhode Island University, has found that a reduction in plankton, which are a source of organic matter 'food' for the nitrogen-removing bacteria, has led to them going on hunger strike and being replaced by less fussy bugs that excrete rather than remove nitrogen compounds. 'The cause may be global warming,' says Fulweiler. 'Plankton blooms are triggered by sunlight. In warmer weather there are more clouds, which could affect the process. Also, warmer temperatures may encourage more grazing by other organisms that consume the plankton.'

Either way, if this continues it may have serious implications for the marine ecosystem, because excess

* *Nature*, vol. 448, 12 July 2007

nitrogen could trigger uncontrolled plant and toxic algal blooms at sea.

MAGNETIC BIOPSY

RESEARCHERS HAVE FOUND a way to facilitate cancer biopsies with the help of a magnet.

At the University of New Mexico and Albuquerque company Senior Scientific*, researchers are testing a new breed of iron oxide-based magnetic nanoparticles that are encased in a biocompatible coating. The coating is 'conjugated' with antibodies that can recognise specific cancer cells. This causes the nanoparticles to stick to the cancer cells, magnetising them. Then, during a biopsy procedure, a magnetic field can be applied to the needle, pulling the magnetic cancer cells onto the needle. Tests in the dish showed that large numbers of cells could be picked up in this way in just a few minutes.

The technique may help to reduce the rate of false negative biopsies, where samples fail to pick up rare cancer cells in some patients. The researchers suggest that it may prove helpful in leukaemias, breast, prostate and ovarian tumours.

Other doctors commenting on the approach have also suggested that it might help to pick up signs of

* *Physics in Medicine and Biology*, vol. 52

'silent' cancer spread into an organ previously believed to be free of disease.

NEUTERED NO MORE

AN AUSTRALIAN COMPANY has recently been granted a European licence to market a system for chemically castrating dogs to spare them the ignominy and discomfort of being neutered*. Peptech, based at Macquarie Park, launched its hormone implant system in Australia in 2004. It's based on two agents called desorelin and suprelorin, which are known as GnRH superagonists. These chemicals prevent the release of the hormones FSH and LH, which are responsible for stimulating the production of sperm and testosterone in the testes.

The Australian implants last about 6 months and cost about $A60–90. The company is now working on a version that will work for 12 months. The major benefit of chemical castration is that it is reversible. 'It gives you the option of breeding later,' says Peptech's Katie Yeates.

---

* www.peptech.com

ASTRONOMY

MOON REALLY IS MADE OF CHEESE

RESEARCHERS IN THE US have got a close look at what they're calling 'the weirdest moon in the solar system', and they've even managed to weigh it! Carolyn Porco, from Boulder, Colorado, writing in *Nature**, has published the first close-up views of Saturn's moon, Hyperion.

They obtained the images using the Cassini spacecraft, which arrived at Saturn in 2004, and it's certainly a strange body. Rather than twirling gracefully around its parent planet like our moon orbits the Earth, Hyperion's an irregular shape and tumbles chaotically through space. But what's really got the space scientists intrigued is its surface, which is peppered with pristine-looking impact craters and more holes than the *Titanic.*

'It looks just like a piece of sponge that someone lifted out of a giant ocean,' says Porco. So why the strange appearance? The team think that it's down to the moon's gravity, or rather lack of it. By measuring the gravitational pull of Hyperion on Cassini as it flew past, they were able to calculate the moon's

---

* *Nature*, vol. 448, 5 July 2007

weight – 5.6 million billion tonnes – and therefore also its density. This showed that, just like Swiss cheese, it is 40% empty space. So when objects collide with the moon and punch a hole, rather than the debris settling out over the surface and giving it a smooth appearance, the low gravity allows the ejecta to disappear into space.

BREATHTAKING NEW DISCOVERY OF ASTHMA GENE

RESEARCHERS IN THE UK have uncovered a gene that triggers asthma. Bill Cookson and colleagues*, from London's Imperial College, compared the genes of 1000 children with asthma and 1000 healthy 'controls' to track down genes that were more common in the asthmatics and might therefore provoke the condition.

To do this the team used a system of genetic markers called SNPs or single nucleotide polymorphisms. These flag certain genetic sequences. By analysing large numbers of people with a disease, and comparing them with people who don't have the condition, you can see SNPs, and hence DNA hotspots, that crop up more often in the diseased individuals than in the healthy ones. Using this technique, the team were able to home in on several DNA hotspots on chromosome 17, and also identify a new gene, called ORMDL3, which was much more common in the children with asthma than the healthy controls.

'This gene occurs in about 30% of children with asthma,' says Cookson. 'It seems to have a fundamental role in the working of the immune system, but

* *Nature*, vol. 448, 26 July 2007

we don't know what it does yet.' So the next step will be to study where in the body it operates and how it works. This could well open up new avenues for the treatment or even prevention of asthma.

But the fact that only 30% of the asthmatic children were carrying it shows that there's much more to asthma than just genetics, and that mystery still needs to be solved.

DISEASES AND DISORDERS

PARKINSON'S REMEDY?

RESEARCHERS FROM FINLAND have uncovered a new growth factor that can promote the survival of the class of nerve cells killed off by Parkinson's disease. Mart Saarma, from the University of Helsinki*, used the power of computers and the internet to trawl through genetic sequences looking for proteins that are secreted by brain cells. The result was CDNF – conserved dopaminergic neurotrophic factor – which is a brain-specific growth factor that powerfully protects dopamine-producing nerve cells.

Rats injected with a neurotoxin that kills these cells could be rescued from developing a Parkinson's-like condition if the newly identified factor was injected even several weeks later. As well as protecting the threatened nerve cells from the effects of the toxin, CDNF could also encourage them to grow additional connections. This suggests that it might help in cases of human Parkinson's by slowing down the progression of the disease, and boosting the performance of the remaining nerve cells in the patient's brains, reducing the symptoms.

---

* *Nature*, vol. 448, 5 July 2007

Although other growth factors for dopamine nerve cells have been uncovered previously, including one called GDNF, these factors have effects in other tissues in the body which have resulted in side effects when they have been used clinically. But CDNF seems to be brain-specific, which might give it the edge.

Saarma is cautiously optimistic. 'The specificity for the brain means that it could be very useful in Parkinson's, but we need to test it first,' he cautions.

GREATEST TITS. BIRDS STAY AHEAD OF THE FASHION WITH CUTTING-EDGE SONGS

SCIENTISTS HAVE SHOWN that birds move with the times by updating their songs; play them an old one and, just like teenagers at a disco, they'll desert the dance floor. Elizabeth Derryberry*, from Duke University, North Carolina, had been studying the process by which birds develop local 'accents'.

Ecologists have suspected for some time that birds have regional dialects and pay more attention to their own dialect than a foreign one, but the rate at which these local languages evolve wasn't known. So Derryberry compared samples of male white-crowned sparrows' songs recorded in 1979 and 2003. The more contemporary song had a lower pitch and was also slower, but would the birds notice? To find out she then played the samples to male and female birds. Just like the Chicken Dance at a wedding, the older material went down like a lead zeppelin with the listening birds, who much preferred the more recent songs. Upon hearing the contemporary material,

* *Evolution*, vol. 61, June 2007

the females solicited more copulations, and the males strutted about oozing territorial aggression.

These results show that, within a relatively short time, meaningful differences in song styles can emerge, and this could have the effect of creating a barrier to mating between isolated populations. Derryberry, who has written up the research in the journal *Evolution*, suggests that this could be one of the ways in which new songbird species emerge.

NON-HUMAN BIOLOGY

SKINCARE BY JELLYFISH

RESEARCHERS IN JAPAN have finally found a use for the huge excess of jellyfish that have been turning up in Japanese waters in recent years – as a source of skincare products.

Kiminori Ushida*, from the Institute of Physical and Chemical Research in Wako, Japan, found that the mucus from five jellyfish species studied was rich in a family of slimy proteins called mucins. These help to lubricate mucosal surfaces such as the front of the eye and the mouth, but they play a big role in cosmetics, where they help to retain moisture, and are also the basis of artificial mucus preparations. But the normal source is not ideal, because they're extracted from cow salivary glands and pig stomachs.

Jellyfish have become a big problem in Japan in recent years, and their increasing numbers have been blamed on over-fishing, climate change and the creation of artificial reefs. They've even blocked up the cooling water intake of a nuclear power station at Hiroshima in the last few years, forcing the power station to temporarily cut its output while the

---

* *Journal of Natural Products* 2007; 70(7); 1089-1092

blockage was cleared. Now maybe what was previously turning into a nuisance, and occasionally a restaurant delicacy, will turn into an asset for the cosmeceuticals industry.

HUMAN BIOLOGY

WHEN SCIENCE IS A YAWN

US RESEARCHERS HAVE come up with a novel theory to explain why we yawn – as a way to cool down the brain! Andrew and Gordon Gallup*, from the State University of New York at Albany, first asked a group of student volunteers to watch a series of video clips of people showing different emotions. In some cases the actors were laughing, in others they displayed neutral expressions and in several cases they were yawning. While they watched, the students were instructed to breathe either through their mouths only, through their noses only, or 'normally'. As they watched the videos, an observer standing behind a one-way mirror counted how many times the student yawned in sympathy with the footage.

Around half of the students breathing only through their mouths or 'naturally' during the video clips were spotted yawning contagiously when the on-screen actor yawned. But intriguingly none of the nose-only breathers yawned. The researchers wondered whether this might be because nasal-breathing was having an effect on the temperature of the brain. To find out, they

---

* *Evolutionary Psychology*, 5(1): 92-101

recruited a second group of students and randomly assigned them to receive either a heat pack (at 46°C), a cold pack (at 4° C) or a pack at room-temperature. They were asked to hold these to their foreheads for 2 minutes before watching the same videos seen by their colleagues.

The results were striking. Virtually no one given a cold-pack yawned, but the yawn rate amongst those with the warm packs was the same as in the previous experiment. The researchers suggest that cooling the head in turn cools the brain (because cold blood flowing out of the head in veins removes heat from blood flowing into the head via the arteries). This is important because fatigue and sleep deprivation are associated with increased brain temperature. Nose-only breathing works similarly because it cools blood in the walls of the nasal passages.

So, say the researchers, contrary to what people think, that yawns are a sign of boredom, they may actually serve to antagonise sleep and maintain alertness. And they're contagious because if someone yawns in a group, showing that they might be nodding off, everyone else yawns to ensure that the rest of the group remain vigilant.

MEDICINE

STEM CELL THAT'S ALMOST HUMAN

SCIENTISTS IN THE UK and America have discovered a new type of mouse stem cell that's effectively identical to human embryonic stem cells. This means that researchers can use these mouse cells to discover how to unleash the power of human stem cells, but without having to work on human embryos. Roger Pedersen from Cambridge University and Ron McKay* from the NIH in Maryland each made the discovery independently. They took cells from a layer in the embryo called the epiblast, which normally goes on to form the future body. When these cells were grown in the dish they behaved identically to human ES cells; they required the same culture conditions to keep them alive and they could turn into all of the cells of the future body. Prior to this finding, scientists had been puzzled for a long time because the cells they collected from mouse embryos and were calling 'mouse ES' cells behaved in a totally different manner to their human equivalent and no one could understand why. It also meant that, despite our thorough knowledge of mouse genetics and development, mice couldn't be used as a

* *Nature*, vol. 448, 12 July 2007

model for the function of human stem cells, because the cells were so different. Now it's clear that scientists were studying the wrong thing. 'What makes these cells so exciting is they provide a model for understanding human embryonic stem cells that adds the awesome power of mouse genetics, and the physiology of rats', Pedersen points out.

NEW ALZHEIMER'S DRUG GOES ON TRIAL

A NEW CLASS of drug designed specifically to combat Alzheimer's disease has entered clinical trials in the US. Dubbed CTS-21166, the drug was the brain child of Purdue researcher Professor Arun Ghosh*. Unlike existing treatments for Alzheimer's, most of which aim to boost levels of the neurotransmitter acetyl choline, CTS-21166 blocks an enzyme called beta-secretase, which is thought to be responsible for the pathological build-up in the brain of aggregates called beta-amyloid. These are a neuropathological hallmark of the disease and are thought to provoke the death and dysfunction of nerve cells, which is what triggers the disease. Scientists suspect that if the formation of these amyloid plaques can be prevented, then the progression or even the onset of the disease might be delayed. The trial, which is being run by San Francisco-based pharmaceutical company CoMentis Inc., has enrolled 48 healthy volunteers in a phase 1 trial to assess the safety and tolerability of the new drug, with phase 2 trials involving patients with Alzheimer's starting in 2008. Ghosh is optimistic. 'The molecule

* news.uns.purdue.edu

is both highly potent and highly selective, meaning it does not appear to affect other enzymes important to brain function or cause harmful side effects'.

NON-HUMAN BIOLOGY

FISH STARVE THEMSELVES TO 'FIT IN'

AUSTRALIAN RESEARCHERS HAVE found that fish operate a strict pecking order that punishes queue jumpers; fear of being shunned by the group leads to some fish starving themselves to avoid confrontations with those higher up the hierarchy. James Cook University researcher Marian Wong* made the discovery by studying the behaviour of gobies living on reefs around Lizard Island on the northern Great Barrier Reef. Among this species, only the top male and female mate. All the other females have to wait their turn in a queue which is based on their size. Each fish has a size difference of about 5% from the one above and the one below it in the pecking order. But if the difference in size decreases below this threshold the smaller fish tries to jump the queue and, responding to the challenge, the superior fish will try to drive the upstart from the group. Expulsion in this way would mean almost certain death so, Dr Wong has found, the fish deliberately stay slimmer than their seniors to avoid rocking the boat!

---

* *Current Biology*, vol. 18, 6 May 2008

DISEASES AND DISORDERS

STEM CELLS GET TO THE HEART OF ANGINA

Researchers in the US have found that patients with angina who received injections of their own stem cells into the diseased heart muscle showed considerable improvements in their symptoms. Cardiologist Douglas Losordo and his colleagues* from the Feinberg Cardiovascular Research Institute, recruited 24 patients aged 48–84 with severe (grade 3 or 4) angina; on enrolment the patients were symptomatic upon mild exertion; even brushing their teeth was sufficient to provoke chest pain. The team harvested stem cells from the patients by injecting them with a hormone called G-CSF. This encourages bone marrow stem cells to grow and spill over into the bloodstream from which they can easily be collected. The team used a molecular marker known as CD34 to single out the stem cells, which were then injected into the heart muscle in some of the patients. Others received a placebo. The injections were placed at sites in the heart known as 'hibernating myocardium', which are regions of the muscle revealed by scans to contain viable muscle which is largely shut down (asleep)

* www.northwestern.edu/newscenter/

due to poor oxygen and glucose supply. Three to six months after the therapy, the patients given stem cell injections had improved significantly. Some of them went from barely being able to make it to the toilet to managing two flights of stairs. The team suspect that the injected stem cells promote the formation of new blood vessels within the muscle, helping to boost the supply of oxygen and therefore increasing the workload that the heart can handle. 'Our goal is to reconstitute the microcirculation, get the blood back into the tissue and alleviate the symptoms,' says Losordo.

MEDICINE

SCIENTISTS TURN RABIES INTO TROJAN HORSE

HARVARD SCIENTISTS HAVE tamed one of nature's nastiest pathogens by turning part of the rabies virus into a powerful therapeutic tool. Manju Swamy and his team* borrowed the surface coat of the virus to produce the molecular equivalent of a Trojan Horse capable of smuggling drugs and other molecules past the 'blood brain barrier' and into the nervous system. The RVG peptide, as it is known, latches onto a chemical docking station called the acetyl choline receptor, which sits on the surface of nerve cells and blood vessel cells that form the blood brain barrier. When it locks on the protein it's taken up into the cells, carrying its cargo with it, which can be used to shut off host genes in nerve cells, or even the genes of infecting viruses like Japanese encephalitis. Using this approach the researchers were able to rescue mice from a potentially fatal brain infection. The team think that their approach might hold the key to highly efficient drug delivery to the nervous system and manipulation of brain neurochemistry in the future.

---

* *Nature*, vol. 448, 5 July 2007

MIRROR MIRROR ON THE MOON

CALIFORNIA-BASED SPACE scientists have come up with a way to make a moon-based telescope 1000 times more powerful than Hubble. But unlike traditional telescopes, theirs use a liquid as its mirror. The idea relies on gravity deforming the liquid, when it is spun, into the perfect mirror shape. But the key breakthough from Roger Angel and his colleagues* at the University of Arizona has been to find an ionic liquid that remains mobile and won't evaporate at ambient temperature and pressure in space, and also the ability to coat the surface of the liquid with silver particles to make it reflective. To obtain the best results the telescope would have to be sited on one of the moon's poles, because the low temperatures there will enable the mirror to detect objects with significant red-shift. Such a powerful telescope would enable us to see further into space, and hence further back in time, than ever before.

---

* *Nature*, vol. 447, 21 June 2007

CHEMISTRY

SUGAR-BASED FUEL PACKS SAME PUNCH AS PETROL

SCIENTISTS HAVE DEVISED a catalytic way to turn the sugar, fructose, into a powerful new fuel, called DMF (dimethyl furan), which could be used as an effective petroleum replacement. The new fuel has several advantages over ethanol, the existing biofuel of choice, including a superior octane rating of 120 (which is excellent), rapid synthesis (it requires no fermentation), low volatility (ethanol evaporates too readily), it doesn't soak up water (and thus rot engines) like ethanol, and it's energetically very favourable to make (ethanol requires energy-hungry distillation to produce). The new fuel, which has been cooked up by Wisconsin-Madison researcher Jim Dumesic*, is made in several steps. First the fructose is heated with sodium chloride (table salt) and hydrochloric acid; this strips the sugar molecule of three of its oxygen atoms by turning them into water, and produces an intermediate called HMF (5-hydroxymethyl-furfural). The HMF is then mixed with hydrogen gas and passed over a metal catalyst containing copper and ruthenium; this removes a further two oxygen atoms, again by producing water,

* *Nature*, vol. 447, 21 June 2007

and leaves DMF. So will this be appearing at the pump before too long? 'That's the hope,' says Dumesic. 'We're planning to start vehicle tests, probably using fuel that is a mixture of DMF and normal petrol to see how it behaves and how clean it is.'

EMERGENCY IMMUNITY

A UK-BASED COMPANY, LifeForce, is offering an 'immune system backup' facility whereby they store white blood cells collected when an individual is healthy so that they can be re-infused to restore the immune system later in life following a disease, such as cancer, treatment for which may require immune-destroying chemotherapy.* Starting with just a small sample of blood, the cells can be numerically expanded in the lab using growth factors that stimulate the production of whole legions of new cells. Using this technique it might also be possible to restore an immunity to a patient whose immune system has been dismantled by HIV. According to LifeForce co-founder Del Delaronde, 'the small blood sample [we collect initially] will have the complete repertoire of all your white blood cells'. So when a patient's immune system fails, even years later, it can be reset using the stored material. 'We send them their pristine system from 25 years ago.'

---

* www.immunesystembank.org/

MEDICINE

HUNTER, AS WELL AS WEAPON

US RESEARCHERS HAVE developed a form of gold-coated glass nanoparticle, which can be used both as a beacon to pinpoint the positions of cancer deposits around the body, and also as a weapon to destroy the cells it's labelling. Size is everything for these particles which, at 140 nanometres across, are just the right dimensions to reflect about 30% of the infrared energy shone upon them, making them visible to a scanner, while turning the remaining 70% into heat that can destroy a cancer cell. The particles are preferentially taken up into cancers because they contain leakier blood vessels than healthy tissue.

Houston-based researcher Andre Gobin and his team*, who are developing the technology, tested the new 'nanoshells' on mice with colon cancer. After an injection with the particles, the animals were scanned using a low-power near-infrared laser. The team measured the amount of light being bounced back from the body. They found that tumour deposits labelled with the particles reflected back 56% more light than normal tissue. They then applied a higher power laser to each

* *Nano Letters*, vol. 7, 2007

of the labelled sites for 3 minutes. Of the mice treated in this way, 80% were still alive 7 weeks later, compared with untreated animals, which all died within 3 weeks.

This means that doctors treating human patients might be able to use this technology to trace the locations of tumours around the body and then, by turning up the power of their light source, zap each of the tumour deposits into oblivion while leaving healthy tissue unharmed.

PHYSICS

RADIATION-HUNGRY FUNGI TO FEED FUTURE ASTRONAUTS

RESEARCHERS HAVE FOUND that the suntan-molecule melanin, which is produced in the skin to soak up UV, is helping fungi to flourish inside the highly radioactive core of the Chernobyl nuclear reactor that exploded over 20 years ago. Writing in *PLoS*, Arturo Casadevall from the Albert Einstein College of Medicine, New York*, said he first came across the phenomenon when he heard how a robot sent into the remains of the Chernobyl reactor had returned with samples of black melanin-rich fungi that were growing inside. Fungi naturally make melanin to protect them from the effects of UV, just like a human, but it looks like they can also use it to obtain energy from radiation. The researchers cultured two species, *Cryptococcus neoformans* and *Wangiella dermatitidis*, and exposed them to radiation counts 500 times higher than background levels; both of them began to grow much faster. The melanin in their cells was capturing the energy of the radiation and locking it into a tame metabolite which the cells could burn. 'Just as the pigment chlorophyll converts sunlight into chemical energy that allows

* *PLoS ONE*, May 2007

green plants to live and grow, our research suggests that melanin can use a different portion of the electromagnetic spectrum – ionising radiation – to benefit the fungi containing it,' said co-researcher Ekaterina Dadachova, who points out that there might be benefits for astronauts too: 'Since ionising radiation is prevalent in outer space, astronauts might be able to rely on fungi as an inexhaustible food source on long missions or for colonising other planets.'

Better hope they like the taste of Marmite (or Vegemite) then.

ASTRONOMY

SEA-CHANGE ON MARS

SCIENTISTS STUDYING THE surface of Mars have come up with the strongest evidence yet that our red neighbour was once home to a giant ocean, which would have covered most of the planet's northern hemisphere. Although a wet history for the red planet has been posited previously, structures resembling shorelines spotted on surveys of the planet varied in height by up to several kilometres, which led some scientists to discount them.

But now, writing in *Nature*, Harvard's Taylor Perron and his colleagues* have shown that these ancient coastlines have been altered by Mars changing its axis of spin, which in turn deformed the surface, explaining why the heights of the shorelines vary so much. The cause, say the scientists, is a phenomenon called 'true polar wander', which we also see here on Earth, where it triggers small annual variations in sea level.

In the case of Mars, at some time in its history the planet spawned the largest volcano in the solar system, Olympus Mons, which is three times the height of Everest. The sheer bulk of this feature caused the spin

* *Nature*, vol. 447, 14 June 2007

axis of Mars to move by 3000 kilometres over the last 2–3 billion years as the planet adjusted its rotation to place the volcano on the equator, which is the most stable arrangement. This caused the crust to buckle, leading to the observed differences in the heights of the shorelines.

But there's another potentially even more exciting spin-off. A planet can only lose water to space at a certain rate, meaning that given the size of the ocean as it once was (roughly Pacific-sized), there must still be significant volumes of water remaining on Mars, most of it below the surface. This makes the existence of potential Martian habitats increasingly likely.

WHEY TO GO – SCIENTISTS MAKE EDIBLE FOOD WRAPPER

Scientists at the US Agricultural Research Service in Wyndmoor, Pennsylvania, have used milk powder and glycerine to produce a water-resistant edible film that could be used to coat or package foods. Peggy Tomasula and her colleagues* have found that high-pressure carbon dioxide can be used as a solvent to extract the protein casein from milk. Mixing this with water and glycerol and then leaving it to dry produces a water-repelling, flexible film-like material which can keep food fresh but is also completely safe to eat and unlike most sandwich packs, completely biodegradable. It would also reassure consumers worried about chemicals leaching from plastic wrappers into foods like cheese, because there's nothing in this film that you wouldn't normally eat anyway.

* www.ars.usda.gov/research/publications/

MEDICINE

DRUG GETS TO THE HEART OF ARTERIAL DISEASE AND DIABETES

US RESEARCHERS HAVE come up with a chemical that can prevent heart disease and diabetes. Writing in *Nature*, Harvard's Gokahn Hotamisligil and his colleagues* have found a way to block a fat-transporting protein called aP2. Under normal circumstances the protein works like a cellular chaperone, guiding fats to different parts of the cell and then funnelling them into different metabolic pathways.

Unfortunately it also assists in helping fats to build up in a harmful way, such as in the walls of blood vessels where they can cause heart disease, or in fat cells where they can trigger diabetes.

But when the researchers used the new drug to block aP2 in experimental mice engineered to have an increased risk of heart disease or diabetes, they found that it significantly cut (by over 50%) the amount by which their blood vessels furred up, and improved their insulin signalling (that is improved their diabetes) compared with untreated controls. Although the team haven't tested the new agent in humans yet, they have found members of the population with defective

* *Nature*, 447, July 2007

aP2 genes. 'These people seem to be protected from heart disease and diabetes, despite having risk factors, suggesting that there's every reason to believe that the agent will work for mice and men,' says Hotamisligil.

HUMAN BIOLOGY

HEY PRESTO, EMBRYONIC STEM CELLS BUT WITHOUT THE EMBRYOS

TWO GROUPS OF scientists have independently stumbled on a way to create tailor-made embryonic stem cells (ES cells), but without the aid of an embryo. The discovery, made by Massachusetts Institute of Technology's Rudolf Jaenisch in Boston US, and Shiya Yamanaka from Kyoto University in Japan*, may hold the key to producing genetically compatible embryonic stem cells for an individual 'on demand', so they can be used to repair damaged tissues and organs.

When a cell becomes specialised to perform a certain task – a process known as differentiation – it loses the ability to carry out other roles or turn into other cell types. This seems to be associated with a switch in the pattern of genes that are turned on in the cell. The key is therefore to find a way to re-program a cell into thinking it's in an embryo again, so that it begins to behave like a stem cell once more.

To do this the teams started with mature skin cells, known as fibroblasts, and used viruses to deliver four essential 'stem cell master genes' known as

* *Nature*, vol. 448, July 2007

Sox-2, c-Myc, Oct4 and Klf4. The results were cells indistinguishable from ES cells. And when the researchers replaced all of the cells of an early animal embryo with the new ES cells they could produce a new animal, proving that they were capable of giving rise to the full range of mature adult tissues.

'This means that we could produce cells specific for an individual that could be used for repair without the risk of rejection,' points out Jaenisch. However, the teams still have to prove that the technique will work with human cells and, as Yamanaka points out, 'The safety of this approach is, at the moment, questionable. We have used retroviruses to introduce the four factors [genes], which could trigger tumours, and one of those factors is a well-known oncogene [a gene that triggers cancer]'. Despite this it's still a massive leap forward and re-programming adult cells to return to an embryonic stem-cell-like state has huge clinical implications.

MEDICINE

MICRO-GYRO MIGHT RESTORE BALANCE TO THE GIDDY

US ENGINEERS HAVE developed an implantable microscopic version of the human balance system, which might help people with damage to their own vestibular systems. Normally we detect movement of the head using a system of three tiny 'semi-circular' canals which are arranged at 90 degrees to each other so they can monitor movement in all three axes. When the head moves, fluid inside the canals pushes on tiny hairs, called cilia, which fire off nerve impulses telling the brain how fast and in what direction the head is moving. But the system can be damaged by injuries, drugs, infections and certain diseases leaving people unable to balance or read and feeling perpetually carsick.

To try to remedy the situation, scientists have tried to produce mechanical replacements for the vestibular system by using a series of micro-gyroscopes locked together at 90 degree angles to mimic the semicircular canals. These gyros consist of a piece of polysilicon that vibrates as a current passes through it. These vibrations are altered, in a measurable way, by rotation in a certain direction.

Unfortunately, the results with these systems have always been far too large to be implanted, making their therapeutic use impractical. But now Andrei Shkel, from the University of California, Irvine*, has found a way to produce a system no larger than 1 cubic millimetre. Even with a battery and processor it would take up no more than 5 cubic millimetres. To fashion the micro-implant he takes a piece a silicon shaped like the 'flat plan' for a cube; on one side he etches a micro-gyroscope, suspended on a spring and supplied by miniature wires laid over the surface of the device. On three other sides he etches accelerometers, which detect acceleration, and then folds up the cube with a needle.

---

* *Audiol Neurotol*, 2006

LEOPARDS LURED BY RINGTONES

FOREST GUARDS IN western India are using mobile ringtones to protect locals by luring leopards lurking near human settlements into traps. Leopards in the State of Gujarat occasionally wander into human settlements looking for food and this can lead to people being attacked and injured. In the past rangers have resorted to tying goats to trees to attract the cats, which then fall into concealed pits. But this can be dangerous and also takes time. So, instead, they're taking a different tack, downloading ringtones of cows mooing and goats bleating. These are played for several hours at a time out of speakers hidden behind a cage. The leopards invariably turn up looking for a ready meal and are safely caught and later released back into forest areas. 'The moos of a cow and bleating of a goat have proved effective to trap leopards,' said one Gujarat forest official. 'This trick works.' Wildlife groups have also welcomed the initiative since it's less likely to result in the captured animal becoming injured.

BUT DOES A HOT MINT STILL TASTE COLD?

US SCIENTISTS HAVE unlocked the secret of how the nervous system senses low temperatures, discovering in the process why sucking a mint makes your mouth feel cold. Writing in *Nature*, David Julius, from the University of California San Francisco*, found that mice lacking a gene called TRPM8 ceased to react to low temperatures. When offered a choice of a warm or cold surface, the TRPM8-deficient mice would spend significantly longer sitting in the cold than normal control animals. The missing gene encodes a pore that sits in the membrane of a specific class of cold-sensitive nerve fibres. When the nerve is cooled down, the pore changes shape, triggering the production of excitatory nerve impulses that are relayed to the brain.

The pore is also activated by menthol, which provokes the pore to change shape at much higher temperatures than it would normally, explaining why sucking a mint makes your breath feel cold against the back of your mouth.

But 'Polos' making your mouth cold isn't the 'hole' story because the mice weren't entirely cold-insensitive.

---

* *Nature*, vol. 448, 12 July 2007

Taking the temperature down below 10°C resulted in an increase in nerve activity, again suggesting that there might be a further 'extreme cold' receptor waiting to be discovered, or that at these low temperatures the physical damage done to tissue is enough to trigger other pain pathways.

Either way, researchers hope that by understanding the workings of these sensory nerve pathways they will be able to design better drugs to block pain syndromes including those triggered by exposure to the cold.

NON-HUMAN BIOLOGY

GOTTA LOTTA BOTTLE

DIET-CONSCIOUS NEW ZEALANDERS may soon be able to tuck into naturally 'skimmed' milk thanks to a program set up to breed a herd of cows that produce milk containing less than a third of the normal levels of fat. Scientists from a Biotech company called Vialactia* discovered a Friesian cow, christened 'Marge', which carries a mutant gene. As a result she produces milk containing only 1% fat, compared with the 3.5% fat normally found in whole milk. Her offspring also produce naturally low-fat milk, indicating that the trait, which the scientists have yet to identify, is dominant.

Another bonus is that the milk also contains high levels of omega-3 fatty acids and makes butter that spreads as easily as margarine even when it's cold. According to Vialactia chief scientist Russell Snell, the company expects to have the first commercial herd of cows supplying naturally low-fat milk and 'ready-spready' butter by 2011.

---

* www.vialactia.com

ASTRONOMY

SOLAR SYSTEM BLOWN INTO EXISTENCE

BY ANALYSING ANCIENT meteorites Danish researchers, reporting recently in the journal *Science*, have re-written the first chapter in the existence of our solar system. The sun and its clutch of planets, including the Earth, date back about 4.5 billion years.

Previously scientists thought that a shockwave from the explosion of a nearby star turning supernova had triggered a ball of gas and other matter to collapse and give birth to a young sun or proto-star. If this were true then remnants from the early solar system, such as meteorites, ought to contain the signature of radio-active iron compounds blasted out from the innards of the exploding star.

But when Martin Bizzaro and his colleagues* from the University of Copenhagen went looking for signs of this iron (in the form of its radioactive breakdown product nickel-60), they couldn't find any. Instead they found another element, aluminium-26. This can only mean one thing: that a supermassive star, dozens of times the size of our own sun and with a lifetime of just a few million years, must have existed in our

* *Science*, vol. 316, no. 5828, May 2007

cosmic backyard at the time the solar system was forming. Stars like this blow off most of their surface layers, including the aluminium-26 they produce, in a million-mile-an-hour maelstrom known as a solar wind. This wind, he thinks, could have buffeted the ball of gas that became us into forming the solar system. '[The absence of any iron-60 signature] rules out the supernova trigger,' says Bizzaro.

IMITATION IS THE HIGHEST FORM OF FLATTERY, AND MAY EVEN SAVE YOUR LIFE

US RESEARCHERS HAVE found that canny moths impersonate the sounds made by their bad-tasting relatives to ward of bat-attacks. Writing in *PNAS*, Jessie Barber, from Wake Forest University*, trained two species of bats to hunt for moths within sight of two infra-red-enabled video cameras. At the same time he recorded the sounds produced by moth prey and the hunting bats. In response to hearing the sonar emitted by bats to locate their meals, some unpalatable tiger moths use a pair of structures called tymbals to broadcast ultrasonic clicks of their own, which are designed to warn the hunter that they taste bad. The bats then duly avoid them. But other tastier moths appear to have muscled in on the trick to avoid being eaten and mimic the bat-repelling sounds with good effect. 'We found that bats do not eat the good-tasting moths that make the similar sounds,' said Barber. Moths that exercised the right to remain silent, on the other hand, fared less well and were instantly gobbled up.

---

* *PNAS*, vol. 104, no. 22, May 2007

METEOROLOGY

RESEARCHERS GET WIND UP ABOUT HURRICANES

US SCIENTISTS HAVE persuaded a muddy lagoon in the Caribbean to surrender 5000 years of hurricane history, enabling them to spot some of the key climate conditions that spawn a fearsome storm.

Writing in *Nature*, Jeff Donnelly and Jonathan Woodruff from the Woods Hole Oceanographic Institute*, report that they collected sediment cores from the lagoons of several Caribbean islands, including one off Puerto Rico called Vieques. Because these lagoons are separated from the sea by a ridge, it takes a big storm, like a hurricane, to push sand and rock grains into them. These grains are then deposited in layers, with each layer corresponding to a different storm from some time back in history.

The team were able to use carbon-dating on the mud mixed with the sand to precisely pinpoint the timings of the storms, going back 5000 years. Then, by marrying up this record with other measures of past climate activity, they were able to show that rising sea temperatures, which were previously thought to be the main drivers of hurricane activity, are not the whole

* *Nature*, vol. 447, 24 May 2007

story. In fact, some of the storms they flushed out in the study were much larger than those occurring today even though the sea was cooler then. A major player, it turns out, is El Niño, which is a pool of warm water that periodically moves eastwards across the Pacific. When this happens it seems to disrupt atmospheric circulation over the tropics, causing developing storm systems to stall in the Atlantic. A strong African monsoon, on the other hand, seems to be linked to more severe hurricane activity. 'So working out what El Niño and the African monsoon are going to do in the future is key to working out what the weather has in store for America,' says Donnelly.

ASTRONOMY

ASTRONOMERS DAZZLED BY NEW SPACE DISCOVERY

It's not just Elton John who's singing about when stars collide, because US astronomers think they've seen it for real. Shri Kulkarni and his colleagues* at the California Institute of Technology were scanning the skies for the signs of supernovae. These are fatal stellar convulsions which culminate in a star blowing itself to pieces in a spectacular light show. Although they didn't spot any, they stumbled upon a strange red glow in the direction of the Virgo cluster. It was about 100 times too dim to be a supernova and about 10 times too bright to be a nova (which is when a star steals some fuel from another star and temporarily flares up brightly). So what was it?

'The explanation which best fits what we've seen is the merger of two stars,' says Kulkarni. 'It surprises people to know that nearly half of the stars in the sky are actually binary systems, with two stars twirling around each other. We think that what we've seen is one such system in which the two stars have merged together.'

---

* *Nature*, vol. 447, 24 May 2007

The result of such a merger is still two stars, but with the two bodies now very close together. 'This can evolve into some very interesting structures, including pulsars and black holes.'

HUMAN BIOLOGY

MIND OVER MAGNET

SCIENTISTS HAVE FOUND that blasting the brain with a magnet might boost mental agility. Fortunato Battaglia, from City University New York*, exposed mice to a procedure called TMS – trans-cranial magnetic stimulation. This is where a powerful magnet is brought close to the scalp and the magnetic field affects the function of groups of nerve cells in the brain. It's used in humans experimentally to explore the actions of different brain areas and also in the treatment of mental illnesses, including depression.

Researchers aren't completely sure how it works. One theory is that it causes long-lasting changes in the sensitivity of the connections between nerve cells, a process called long-term potentiation. This was indeed what the mice showed after five TMS treatments; their nerve cells were more responsive to the excitatory nerve transmitter glutamate.

There was a far more exciting finding when the researchers examined the animals' brains. Among the treated animals there were significant increases in the proliferation of stem cells in the hippocampus,

* www.abstracts2view.com/aan2007boston/

the part of the brain concerned with learning and memory. This suggests that TMS in some way encourages the growth and survival of these new nerve cells which are being born in this part of the brain. Previously, only antidepressants, sex, and physical activity had been shown to do this, meaning that TMS might have a role to play in the prevention of dementia among the elderly.

HUMAN BIOLOGY

IT'S ALL IN THE WRIST, OR SHOULD THAT BE INDEX FINGER?

A STUDY OF 75 school children has shown that a photocopy of their hands is all that's needed to predict their literacy and mathematical skills. Dr Mark Brosnan, from Bath University*, measured the lengths of the index and ring fingers from images of the childrens' hands and compared the results to their SAT (standardised assessment test) scores. The results showed that the greater the length of the index finger relative to the ring finger, the better the subjects performed in maths, and the lower the ratio, the better their literacy skills.

Similar relationships have been found in the past for sporting ability, and scientists think that the relative lengths of the digits reflect the levels of hormones to which a developing baby is exposed in the uterus. The more testosterone there is floating around, the longer the index finger, and the team also think that testosterone could promote the development of brain areas linked to numerical and spatial skills. Oestrogen, on the other 'hand', is thought to do the same thing for those parts of the brain concerned with verbal ability. 'We're not suggesting that finger length measurements

* *British Journal of Psychology*, 99

could replace SATs,' says Brosnan. 'But finger ratio provides us with an interesting insight into our innate abilities in key cognitive areas.'

VIROLOGY

THE VIRUS THAT CAME IN FROM THE COLD

US RESEARCHERS HAVE performed the microbiological equivalent of fighting fire with fire by showing that animals infected with members of the herpes virus family are much better at fighting off infections caused by virulent bacteria than uninfected animals.

Reported in *Nature*, Skip Virgin and his team at Washington University in St Louis* infected mice with one of two herpes viruses, either MHV-68 or mouse CMV, both of which cause the rodent equivalent of human glandular-fever-like illnesses. After the initial infection had passed, the animals were then challenged with potentially lethal doses of either Listeria, or a second bug known as *Yersinia pestis*, which causes bubonic plague.

Compared with a group of control animals, the virally infected mice were all highly resistant to the bacteria and the team found higher levels of at least two immune-boosting hormones, called interferon-gamma (IFNg) and TNF-alpha, in their bloodstreams. 'This suggests that the viral infection is in some way beneficially modifying the immune system of the host,' says

* *Nature*, vol. 447, 17 May 2007

Virgin. 'A key feature of the herpes viruses is that, after infection, they remain in the body for the lifetime of the host, a state known as latency. It therefore stands to reason that the virus should want to protect its host, because if the host dies the virus dies.'

As yet the team are unsure if the effect also exists in humans, and they're now exploring whether the protective effect extends to defending animals against other viruses and whether it will be possible to recreate the effect without having to be infected with the virus – perhaps by means of some kind of vaccine.

ASTRONOMY

MOON, RUBBED UP THE WRONG WAY, LETS OFF STEAM

SCIENTISTS HAVE SOLVED the mystery of a massive geyser seen spurting from the surface of one of Saturn's moons.

In late 2005, NASA's Cassini space probe picked up a strange hot-spot on the south pole of Enceladus, one of Saturn's many moons. Closer inspection revealed a huge plume of hot water 500–1000 kilometres tall that periodically jetted out from the moon's surface. This Enceladian 'old faithful' is so large that the material it ejects even makes it into space where it contributes to the rings of Saturn.

This phenomenon left scientists with a big puzzle to solve. At only 500 kilometres across, Enceladus is a tiny moon resembling an ice-bound marble. It would have cooled down long ago, so where was the energy coming from to produce such a outburst?

Two papers in *Nature*, by NASA's Terry Hurford and UC Santa Cruz researcher Francis Nimmo*, shed some light on the mechanics of Enceladus's hot bottom end. The orbit of the moon around Saturn is not circular, it's an ellipse, so the gravitational pull

---

* *Nature*, vol. 447, 17 May 2007

exerted by the planet changes along the way. This squeezes and pulls the surface material of the moon, generating heat and opening up cracks in the ice. These weak-spots then allow the heated material to escape. 'But for our calculations to work there must also be a big ocean beneath the icy surface of Enceladus,' points out Nimmo. 'If the ice were sitting just on a rocky surface there wouldn't be much movement, because that's stiff. But with a big ocean beneath the surface you could get big tidal movements that would generate lots of movement in the surface and create the necessary heating effect. It's unleashing the energy-equivalent of a few hundred power stations.'

HUMAN BIOLOGY

HAIR TODAY, GONE TOMORROW . . . OR PERHAIRPS NOT

SCIENTISTS HAVE CHALLENGED the dogma that mammals cannot regenerate lost organs, and also given hope to comb-over fans everywhere, by showing that mice can recover lost hair follicles.

University of Pennsylvania researcher George Cotsarelis and his colleagues* made full-thickness skin wounds in mice to study the healing process. But they were surprised to see that in animals with fairly large wounds the mice didn't heal with a bald patch; instead they grew new hair follicles.

To find out where these new hair-producing organs were coming from the team used a genetic labelling technique to track cells that moved into the wound from its margins. They expected adjacent hair follicles to be contributing some kind of stem cell, but instead the new follicles were being produced by a population of skin cells from the wound edges. These skin cells seemed to have re-activated an embryonic pattern of gene expression which would have produced hair follicles when the animal was first developing.

To confirm that this was the case, when they

---

* *Nature*, vol. 447, 17 May 2007

blocked the action of one of the genes involved in this pathway the researchers were able to prevent any new follicles from forming, and when they over-expressed the same gene, called wnt, they were able to increase the birth rate of new follicles. This finding suggests that it should be possible to switch on these stem cells to achieve the same feat in people affected by scarring, alopecia and male pattern baldness. So there's hope for me yet!

CHEMISTRY

AL-UMINATING ANSWER TO FUEL CELL PROBLEM

US SCIENTISTS HAVE stumbled on a safe way to make large amounts of hydrogen, on demand, to fuel environmentally friendly vehicles.

Jerry Woodall, from Purdue University*, has found that an alloy made by mixing gallium and aluminium can be reacted with water to produce large amounts of hydrogen gas, which can be collected and fed to an engine.

During the reaction only the aluminium is used up; it turns into aluminium oxide. This can be recycled together with the gallium, which is itself not consumed by the process. The trick works because normally aluminium won't willingly react with water because the metal forms a protective oxide 'skin' on its surface. But the gallium interferes with the formation of the oxide layer, enabling the aluminium to react freely with the water. The two metals are melted together to produce the alloy, which is then turned into pellets for the reaction with water.

The benefit of hydrogen-powered vehicles is that they are very clean since the only exhaust product is

* news.uns.purdue.edu

water. And if energy used to recycle the aluminium oxide back to aluminium came from non-fossil sources, the process could even be carbon-neutral, and feasible too. 'A mid-sized car with a full tank of aluminium-gallium pellets could take a 350 mile trip and it would cost US$60,' Woodall points out.

ASTRONOMY

WEATHER FORECAST THAT'S OUT OF THIS WORLD

A TEAM OF Harvard-based researchers have produced the first example of a weather forecast for a planet outside the solar system.

Writing in *Nature*, Heather Knutson and her colleagues* describe how they have used NASA's Spitzer Space Telescope to watch a Jupiter-sized planet, which is catchily named HD 189733b, orbiting its parent star 63 light years away.

Unlike our own planet Jupiter, which takes 12 years to complete a lap of the solar system, this distant gas giant sits very close to its parent star and orbits every two days. But a consequence of being so close to the star is that the planet has become tidally locked, meaning that, just as our moon always shows us the same face, HD 189733b always points the same part of its surface at the star and has a dark-side looking towards outer space. By carefully measuring the colour of the planet as it moved from being in front of the star (showing its dark side) to adjacent to the star (and showings its lit side) the team were able to calculate how hot it must be, because the brighter something

* *Nature*, vol. 447, 10 May 2007

is the hotter it usually is. The researchers were able to calculate that the side facing the star was a scorching 1200° Kelvin, but the real surprise was that the dark side was nearly as hot at over 900° Kelvin.

The team think that strong winds carry the heat from the lit side round to the dark side, because the hottest spot on the surface of the lit side lies in a direction downwind of the point on the planet's surface closest to the star. 'This is the first time we've got a glimpse of the climate of another world in another solar system,' says Knutson.

GENETICS

WHY RARITY IS VALUABLE

A CANADIAN SCIENTIST has confirmed what we suspected all along – that there is something special about being one in a million, at least when it comes to genetics.

Writing in *Nature*, Toronto University's Marla Sokolowski* has uncovered the mechanism that keeps different versions of a gene in play in the game of life. Using fruit fly larvae she identified two genetic variants with markedly different behaviours known as 'sitters' and 'rovers'. Sitters do just that, they sit and eat, while rovers are more prone to wander off and explore new ground. These differences are controlled by two different versions of a single gene, called the foraging gene, making the trait relatively easy to study. The larvae were reared so there were either equal numbers of sitters and rovers, sitters outnumbered rovers three-to-one, or rovers outnumbered sitters three-to-one. In all cases, larvae carrying the form of the gene that was rarer did better. Why? Because, if you are a rover surrounded by a glut of immobile sitters and you wander off, you are much more likely to find food than if you stay put and try to compete with the hungry horde. Conversely, if

* *Nature*, vol. 447, 10 May 2007

you are a lone sitter amongst a population of highly mobile rovers and you find a good food source, by staying put you can make the most of it; the rovers meanwhile wander off and miss out.

This finding helps to explain why we see the combinations of genes that we do in nature, but the team still need to prove that it happens in the wild rather than just in the laboratory. The next step will be to test other species including bees, mice and humans, since even we have a version of the foraging gene.

POLLUTION BREAKTHROUGH LEAVES RESEARCHERS BREATHLESS

RESEARCHERS AT THE University of North Carolina have uncovered a possible link between air pollution, heart disease and allergies. Dr Karin Yeatts and her colleagues* followed up 12 adult asthmatics aged 21–50 over a period of up to 11 weeks. Among other data collected by the team, blood cell counts, lipid levels and measurements of heart rate were taken on nine occasions from each of the subjects, and at the same time measurements of air quality including particulate matter were made.

The results showed that an increase in airborne particulates of one microgram per cubic metre provoked a 5% increase in blood triglycerides (lipids), a 0.16% increase in eosinophils, a form of white blood cell linked to allergy, and a decrease in heart rate variability, which can be a sign of cardiac stress. These results are intriguing because they confirm previous reports of increased numbers of heart attacks on high-pollution days. They also agree with a recent study, which saw a group of elderly rats go on a 6-hour road-trip around the state of New York, during which

---

* *Environmental Health Perspectives*, May 2007

they breathed 'road air'; the animals returned with similar reductions in their heart-rate variability.

At the moment researchers don't know how pollution provokes these changes, but steps to ensure that risks are minimised will certainly be a breath of fresh air for those living next to a busy road.

CHEMISTRY

SMELLING YEAST AN EXPLOSIVE COMBINATION

SCIENTISTS HAVE MADE it possible for yeast cells to sniff out explosives. Writing in *Nature Chemical Biology*, Danny Dhanasekaran and colleagues, from Temple University School of Medicine in Philadelphia*, describe how they 'borrowed' the chemical smelling system from a rat and successfully inserted it into a yeast cell. They located a receptor (a chemical docking station) that recognises a breakdown product (known as DNT) of the explosive TNT. This receptor was coupled on the surface of the yeast cell to a signalling molecule which could switch on a green coloured gene whenever the explosive chemical was picked up by the receptor. As a result, the cells would turn themselves green whenever they were exposed to TNT.

This is just proof of principle, because the same trick could be used to pick up a range of important chemicals. Detectors based on yeast like this would also be a fraction of the size of the man-made equivalent. 'This could be of immense value in the detection of environmental toxins and chemical warfare agents, even at sub-lethal levels,' said Dhanasekaran.

---

* *Nature Chemical Biology*, vol. 3, 2007

SCIENTISTS SWOLLEN WITH SUCCESS

RESEARCHERS HAVE UNCOVERED a natural Viagra-like chemical in the venom of a Brazilian 'wandering' spider, *Phoneutria nigriventer*. Kenia Pedrosa Nunes, Romulo Leite and colleagues, from the Medical College of Georgia*, followed up on anecdotal reports that male victims bitten by the spiders subsequently developed a sustained erection. Analysing the components of the venom one by one, they uncovered a small protein, known as Tx2-6, capable of recreating the effect in male rats. Tests on the animals show that the protein works in a novel way compared with drugs like Viagra; it boosts the levels of an important vascular signalling molecule called NO (nitric oxide), which causes bloods vessels to relax and open up. It might therefore hold the key to managing a range of vascular disorders as well as impotence. The scientists also suggest that if given with other drugs, like Viagra, it could help to magnify their effect. According to Leite, 'the combination of the two drugs could be even more efficient in patients that don't respond well to Viagra'.

---

* www.scielo.br/

HOME SWEET REEF

WHERE DO THE FISH swimming around on remote reefs come from and how do they get there? Thanks to researchers in Australia, together with some willing fish, we now know the answer.

Working on a reef in Kimbe Bay, Papua New Guinea, scientist Geoffrey Jones and his colleagues* from the James Cook University in Queensland first netted female coral fish from a 0.3 square kilometre region of the reef. The fish were injected with a stable but rare isotope of barium, which is passed into the fishes' eggs and subsequently into the skeletons of their offspring, giving the babies a characteristic chemical signature. Following the injections, the fish were released back into the sea to breed.

A few weeks later the team returned to the reef and collected young fish to test them to see if they carried a barium 'tag'. The scientists were surprised to see that over 60% of the fry they picked up were positive. This is intriguing because, as Geoffrey Jones points out, 'marine fish lay very small eggs, and when they do, they are released into the water column. They develop into tiny little larvae that we think drift around in

* *Science*, vol. 316, no. 5825, 4 May 2007

water currents, sometimes for months.'

But the missing part of the puzzle was always where do these larvae go? Now we know that the answer is that, more than half the time, they head back to the reef that spawned them. These results are important because they demonstrate why marine reserves – conserved areas where fishing and other forms of exploitation are forbidden – are a good way to protect rare and threatened species, because sufficient numbers of young should return to the protected area to sustain numbers over time.

The big outstanding question that the scientists still have to solve is how the fish found their way home in the first place, but that's another 'dory'!

HEAL THYSELF OF HIV

GERMAN SCIENTISTS HAVE found the tail-end of a protein naturally present in the human bloodstream could hold the key to a new generation of anti-AIDS drugs.

Writing in the journal *Cell*, University of Ulm researcher Jan Munch and colleagues* have discovered that a small fragment of a larger blood protein, called alpha-1 anti-trypsin, can block the ability of HIV to invade susceptible immune cells.

The new factor, which they are calling VIRIP, short for virus inhibitory peptide, consists of the first twenty amino acids from the parent protein. When added to cells in the dish, VIRIP protected them from infection with HIV. The molecule seems to work by locking onto a structure on the viral surface called gp41, which is normally kept hidden until the moment when the virus tries to invade a cell. This so-called fusion peptide causes the surface of the cell to merge with the virus, enabling it to enter.

Blocking this key protein keeps the virus out and because gp41 is essential for the virus to be able to

---

* *Cell*, vol. 129, 20 April 2007

infect a cell, unlike other anti-HIV drugs the virus struggles to develop resistance to the action of VIRIP because to do so compromises infectivity. The agent also worked against drug-resistant strains of the virus and the team have found that by substituting certain chemical groups in the VIRIP molecule they can boost its protective power over 100-fold.

So why are humans still susceptible to HIV if we carry this molecule naturally? Probably because there isn't enough of it around to prevent infection with the virus in the first place, but if some people have more VIRIP in their bloodstreams than others this may explain why they are less vulnerable to HIV infection, or progress to AIDS more slowly than patients with lower levels. 'The findings reveal a new target for inhibiting HIV that remains fully active against viral strains that are resistant to other drugs,' said Frank Kirchhoff, one of the study authors. 'That's a big advantage.'

ACKNOWLEDGEMENTS

This book would not have been possible without the help of a few very fine people.

Firstly my family, without whom I'd be nowhere, my friends at ABC Radio National, including Lynne Malcolm, Robyn Williams and the breakfast team who got me delving into the literature in the first place, and The Naked Scientist's crew at Cambridge University – Ben Valsler, Diana O'Carroll, Meera Senthilingam and Dave Ansell, who make it all possible and incredibly good fun. Thank you all.

INDEX

Acacia drepanolobium (tree) 33–4
acetic acid 81
Acropora millepora (coral) 101
Adelie penguins 162–3
adhesives 192
adipocytes 100
adolescence, smoking during 37–8
Africa and human evolution 138–9, 169, 196-7
AIDS 135, 178–9, 281–2
air-conditioning, alternative methods 148–9
air pollution 41–2, 140–1, 275
airport security scanners 57
alcohol, effects on health 31
Ald (gene) 166
algal blooms 206–7
allergies, link to pollution 275–6
allinase (enzyme) 15
Alzheimer's disease 7–8, 12, 42, 70, 87–8, 225–6
  link with herpes virus 87–8
alpha-interferon (hormone)
Amazon 25
amino acid sulphoxides 15
amputation and amputees 55, 73–4
amyloid plaques 7–8, 70, 87–8, 225–6
anaesthesia 115–16
ancient humans 138
androstenone 136
angina 228–9
animal biology and behaviour 25–6, 29–30, 33–4, 47–8, 61–2, 65–6, 81, 101–4, 111–14, 119–20, 156–7, 192–3, 217–18, 219–20, 227, 248, 251, 254, 279–80
animal cloning 68
animal experimentation *see* macaques, mice, rats
anorexia 128–9
Antarctic 162–3
anterior cingulate cortex (ACC) 13
anthropology 196–7
antibiotics 93–4
antibodies 208–9
ants 25–6, 111–12
ApoE4 (gene) 87

Index

Apophis (asteroid) 172
appetite-stimulating hormone 1–2, 76
arachidonic acid 79
Arenicola marina (lugworm) 90
art, use in studying past climates 121–2
arterial disease 185–6, 242–3
artificial sweeteners 170
asteroids 152–3, 172–3
asthma 213–14
astronomy 83–4, 172–3, 200–201, 211–12, 231, 239–40, 252–3, 257–8, 265–6, 271–2
attractiveness, female, link to oral contraceptive pill 109–10

b12 (antibody) 135
bacteria 59–60, 71–2, 81, 85–6, 206–7
  E. coli 85, 93–4
bacteriophages 85
bad breath 59–60
balance, sense of 246–7
Baptistina parent body (asteroid) 152–3
bats 254
batteries, lithium 27–8
bears, grizzly 113–14
bees 61–2
beta-amyloids (proteins) 7, 87–8, 225–6
beta-endorphin 17
beta-glucuronidase (GUS) (gene) 166
beta-myrcene 118
beta-secretase (enzyme) 225–6
bicycles 181
bioaccumulation 162
biofuels 232–3
biology, human *see* human biology
biopsies 208–9
bipolar disorder 12
bird flu 126–7
birds 26, 29, 47–8, 217–18
bird 'accents' 217–18
bleeding disorders 134
blood clotting 134, link to pollution 140–1
blood platelets 134
blood sugar levels 43–4
blood transfusions 85
body clock 11–12
bone marrow 19, 100
brain cells, link to diabetes 158–9
brain damage 42
brain development 37, 41–2
brain, flow of information to 38
brain lesions *see* amyloid plaques
brain scanning 13
brain temperature, link to fatigue 221–2
breast cancer 5–6, 50
breast milk and breastfeeding 79, 154–5
British Isles, geological formation of 190–1
broccoli 97–8

brown clouds 176–7
bubonic plague 263

caffeine 21–2, 123
cancer, breast 5–6, 50
cancer, 18, 57, 69–70, 124–5, 184, 208–9, 235–6
skin 97–8
Cancer (constellation) 83
capsaicin 115–16
carbon-dating 255
carbon dioxide (CO2) 72, 160–1, 176–7, 189
carcinogens 98, 132
cars, environmentally friendly 3
Cassini spacecraft 211–12, 265
castration, chemical 210
Cephalotes atratus (ant) 25
chemistry 27–8, 123, 148–9, 164–5, 232–3, 269–70, 277
chemotherapy 144, 234
chewing gum 59–60, 142–3
chilli, role in anaesthesia and pain relief 115–16
chimpanzees 9–10
chlorhexidine 60
chlorine 60
chromometer 97
chromosome 11
circadian rhythms 12
cirrhosis 105
climate and climatology 121–2, 160–1, 219
clodronate (drug) 140
cloned animals 68
cloning of stem cells 67–8
clotting, blood 134
coagulation factors 140–1
CO2 *see* carbon dioxide
cold sores 87
common cold 9
computer technology 49–50, 61–2
connexin 43 (protein) 46
conserved dopaminergic neurotrophic factor (CDNF) 215–16
coral 101–2
cortexes, brain 13, 14, 128–9
cosmetics, use of jellyfish in 219–20
cows 251
Crematogaster mimosae (ant) 34
Crematogaster sjostedti (ant) 34
chromosome 17
CRY (cryptochrome) 101–2
Cryptococcus neoformans (fungus) 237–8
cycads 117–18
cytokinin (CK) (gene) 54

DDT 89
'Delhi belly' 130–1
dementia 70, 87, 260
depression 12, 259
diabetes 43–4, 158–9, 194–5, 242–3
diagnostic devices and techniques 57–8, 126–7, 165–5
dialysis 20

di-hydroxy phenylalanine (DOPA) 103, 193
dinosaurs, extinction of 152–3
disease, link to loneliness 150–1
disease, molecular links to 49–50
diseases and disorders 5–6, 7–8, 9–10, 43–4, 87–8, 105–6, 124–31, 158–9, 194–5, 213–14, 215–16, 225–6, 228–9, 275–6, 281–2
disinfection techniques 86
DNA 57, 65, 67, 69
DNA hotspots 213
  coral 101
docosahexaeonic acids 79
dogs, effects of pollution on 42
Dolly (cloned sheep) 68
dopamine 103
driving, dangerous 180
drought and plants 53–4
drugs 178, 225–6, 281–2
  drug delivery systems 144–5, 230
  drug-resistant pathogens 86, 131
  drug tests and trials 49–50
  drugs, anti-AIDS 281–2
  drugs, anti-cancer 144–5
  drugs, immunosuppressive 19–20
dust, link with sea temperatures 39–40
E. coli bacteria 85, 93–4
eardrums, burst 202–3
ear-pinning techniques 146–7
Earth 83–4
ecology 41–2, 51–2, 89–90, 117–18, 132–3, 160–1, 162–3, 176–7, 206–7
ecosystem, marine 206–7
eczema 14, 183
edelweiss 91–2
EDF (molecule) 94
eels, moray 156–7
El Niño 174, 256
electricity 3, 202–3
electrodes 27–28
electromagnetic radiation/ waves 58
elephants 33–4
embryonic stem cells 67–8
emphysematous change in lungs 64
Enceladus 265–6
energy 3–4
energy, alternative sources 3–4, 269–70
environmental pollutants 89–90
enzymes 15, 71–2
embryonic/foetal development 21, 125, 187
emotions 13
erections, sustained 278
Ethiopia and human evolution 138–9
eumelanin (skin pigment molecule) 95
evolution 138–9, 168–9, 196–7

exercise, effects on health 31–2
exhaust fumes 41–2
explosives 277
extinction, effects of simulated 33–4
extinctions, mass 152–3
extraterrestrial life, potential 84

55 Cancri (star) 83
FADS2 (gene) 79
fat metabolism 198–9
fatty acids 79
fibrinogen (clotting chemical) 140
fibroblasts 12
finger length ratios, link to literacy and numeracy skills 261–2
fish 227, 279–80
flavour intensifiers 143
floods, link to sunspots 174–5
flu 126–7
foetal/embryonic development 21, 125
follicles, hair, regenerating 267–8
food 1–2, 15–16, 97–8, 142–3, 162–3, 170–1, 241
food chain 89–90
food cravings 1–2
food technology 241
foot comfort (shoes) 80
forecasting volcanic eruptions 23–4
fossils 168
FoxP2 (gene) 47
fruit flies 1, 204, 273
fuel technology 232–3, 269–70
fumaroles 72
fungi, effect of radiation on 237–8

gastric upsets 130–1
G-SCF (hormone) 228
geckos 192
'gene silencing' 16
gene, 'foraging' 273–4
gene 'modules' 93
generator, knee-mounted 3–4
genes 79, 87, 98, 124–5, 136–7, 150–1, 166–7, 185, 194–5, 204–5, 213–14
genes, anticancer 69–70
genes, marker 12, 194
genes, parents' 69
genetic diversity 196–7
genetic modification (GM) in plants 15, 53–4
genetic transfers 67–8
genetics 53–4, 67–70, 95–6, 101–2, 204–5, 223–4, 273–4
genome, hepatitis C 106
genome, human 50
genome, Neanderthal 95
geology 23–4, 121–2, 152–3, 190–1
geothermal regions 71–2
ghrelin (hormone) 1–2, 76
ghrelin O-acyltransferase (GOAT) 1
Giardia lamblia 130–1

giraffes 33
glacial retreat 177
global warming 72, 176–7, 189, 206–7
glue, mussel 103–4
GnRH superagonists 210
gravity/gravitational effects 83, 152–3, 211–12, 265–6
Great Barrier Reef 101
green fluorescent protein (GFP) 164
Greenpeace (India) 132
GSTA1 (gene) 98

haem oxygenase 1 (gene) 98
hair, regeneration 267–8
hair, use in breast cancer screening 5–6
halitosis 59–60
HD189733b (planet) 271–2
heart attacks 45–6, 140–1, 185, 275–6
heart disease/problems 50, 228–9, 275–6
heart, growth of new 35–6
heavy metals 132
helium-3 63
hepatitis C 105–6
herpes virus 87–8, 263–4
  link with Alzheimer's disease 87–8
Hindus 132–3
hippocampus 77, 259
HIV/AIDS 135, 178–9, 234, 281–2
hominids 138–9
Homo erectus 138–9, 168–9
Homo habilis 138, 168–9
Homo sapiens 168
honokiol 59
hormones, immune-boosting 263–4
hormones, immune-regulating 105
hormone, immune-signalling 140
hormones, role in appetite suppression 76
hormones, role in sleep deprivation 76
HoxD10 (gene) 125
human biology 11–14, 17–22, 31–2, 35–8, 45–6, 59–60, 63–4, 75–6, 77–8, 79–80, 99–100, 115–16, 134, 140–1, 150–1, 154–5, 183–4, 185–6, 198–9, 221–2, 244–5, 249–50, 259–60, 261–2, 278
human genome 50
human metapneumovirus (HPMV) 9
human origins 168–9, 196–7
human speech 47–8
humans, ancient 138, 168–9, 191
hurricanes, formation of 39–40, 255–6
hydrogen, production of 269–70
Hyperion 211
hypothalamus 11, 158–9

identical twins 69–70

IL6 (hormone) 140–1
immune system 135, 234
immune systems in mice 65–6
immunosuppressed people 98
immunosuppressive drugs 19–20
implants 104, 146–7, 246–7
impotence 278
impulsive stimulated Raman scattering (ISRS) 85–6
insects 25–6, 33–4, 117–18
insula cortex 14, 128–9
insulin levels 44
integrase (protein) 178
interbreeding 65–6
interdependence of species 33–4
interferon therapy, side effects of 106
Internet 62
iPods, lightning injuries when wearing 202–3
IQ, link to breast-feeding 79
IQ, link to pollution 41–2
isopentenyltransferase (IPT) (gene) 53
isotopes, chemical 119
itching and scratching 13–14, 183–4
IVF 187–8

jaws, eel 156–7
jellyfish 77, 164, 219–20
jetlag 11
Josephson junction 58
Jupiter 83

kidney transplants 19–20
Krakatoa 121

lacrimatory factor synthase 15
language learning and production 47–8
lapdancers 109–10
lasers 85
leaf senescence 53
leopards 248
leptin (hormone) 76
lightning strikes 202–3
Listeria 263
literacy skills 261–2
loneliness, link to health problems 150–1
longevity, factors affecting 31–2
long-horned beetles 34
lugworms 90
lumbar puncture 17
lung damage due to smoking 63–4
lung function, smokers vs non–smokers 63–4
lysine 103

macaques, experimental 135
macrophages 140–1
magnetic resonance imaging (MRI) 63–4
magnets, use in biopsies 208–9
magnets, use in brain stimulation 259–60
magnolia bark 59–60
magnolol 59

major urinary proteins (MUPs) 65–6
maraviroc (drug) 178
marine biology 101–2, 227, 279–80
marine reserves 279–80
marker genes 12
Mars 239–40
materials technology 148–9, 192–3
MazE (toxin) 93
mazEF (gene 'module') 93
MazF (toxin) 93
medical technology 73–4, 187–8
medicine 55–6, 85–6, 109–10, 146–7, 208–9, 210, 223–4, 230, 234–6, 242–3, 246–7
melanin 237–8
melanocortin 1 receptor gene (mc1r) 95–6
melanocyte stimulating hormone (MSH) 95
memory formation and loss 77–8 *see also* dementia
menstrual cycle 109–10
menthol 249–50
metabolic response 198–9
metal plating 104
meteor showers 153
meteorology 39–40, 174–5, 255–6
methane 71–2
methane monooxygenase (MM) (enzyme) 71–2
methanol 72
mice, experimental 7, 45, 65–6, 77–8, 99–100, 124, 134, 140, 158–9, 183–5, 188, 223, 235–6, 242–3, 249–50, 259, 263–4, 267–8
micro RNAs 106, 124–5
microbes *see* bacteria
microglia 8
micro-plastics 89
microwaves 57
milk, naturally low-fat 251
mimicry, animal 254
mirrors, use of in pain relief 55–6
mirrors, use of in space telescopes 231
miscarriage 21
mobile phones 3, 248
Moon, 265–6
moonlight, link to coral spawning 101–2
moose 113–14
moray eels 156–7
morphine 17
moths 254
mucins 219–20
multiphoton microscopy 7
music, effect on plants 166–7
mussels 103–4, 192–3
Myrmica scabrinodis (ant) 111–12

nanoparticles 144–5, 208–9, 235–6
nanorods 192

nanowires 27–8
  solar 107
Neanderthals 95–6
near-Earth objects (NEOs)
  172–3
nematode worms 25–6
nerve cell degeneration 77–8
nerve cells, 11, 259–60
nervous system 37–8
neural activity patterns 128–9
neutering (dogs) 210
nicotine 38, 154–5
nitrogen-15 (isotope)
noise *see* sound
nonylphenol 89
NQO1 (gene) 98
numeracy skills 261

obesity 1–2, 170–1
obesity control 100
obesity, link to diabetes 44,
  158–9
obesity, link to poor sleep
  75–6
ocean temperatures, link to
  weather 39
oestrogen 261
'oestrus', female/human
  109–10
Olympus Mons 239
onions 15–16
opioid painkillers 18
OR7D4 (gene) 136
oral contraceptive pill 109–10
organ decellurisation 35
organ transplants 19–20
organs, growth of new 35–6
ORMDL3 (gene) 213–14
overeating behaviours 170–1
ozone 189

packaging technology 241
paclitaxel (drug) 144–5
pain and pain control 17–18,
  55–6, 115–16, 250
pain, sense of in prawns/
  crustaceans 81
paintings, use in
  reconstructing climate
  121–2
parasites, ant 25–6
parasites, human 130–1
Parkinson's disease 215–16
particulates 41–2, 140–1,
  275–6
passive smoking 63–4
PCBs 89
'pecking order', in ants 111–12
'pecking order', in fish 227
penguins 162–3
phaeomelanin (skin pigment
  molecule) 95
phagocytes 140
phantom pain 55–6
phenanthrene 89–90
Phoneutria nigriventer (spider)
  278
physics 57–8, 172–3, 237–8
pigment molecules (skin) 95
pill, oral contraceptive 109–10
planetary rotation 340
planetary systems, other than
  ours 83–4
plankton 206–7

plant biology and behaviour 53–4, 91–2, 117–18, 160–1, 166–7, 189, 237–8
plants, high-altitude 91–2
plastics, biodegradable 51–2
plastics, electrically conductive 104
plastics, polluting 89–90
platelets, blood 134
pneumonia 10
poisons, organic 89–90
pollen and pollination 117–18
pollution and pollutants 41–2, 89–90, 132–3, 140–1, 162–3, 176–7, 185–6, 189, 206–7, 275–6
poly 3-hydroxy-butyrate (PHB) 51
polydopamine coatings 104
polymerase chain reaction (PCR) 127
polymers 51, 103–4, 142–3, 192–3
prawns, ability to feel pain 81
prefrontal cortex 14
pregnancy, caffeine during 21–2
pregnancy, smoking during 37
prostheses and prosthetics 73–4
protein plaques *see* amyloid plaques

QX–314 (anaesthetic derivative) 115

rabies 230
radiation, effect on fungi 237–8
radiation, electromagnetic 57–8
radio waves 57
rainfall, heavy, problems caused by 174–5
raltegravir (drug) 178
rats, experimental 17, 35, 115, 170, 215, 278
rbcS (gene) 166
red-headedness 95
rejection, of transplanted organs 19–20, 36, 245
respiratory syncytial virus (RSV) 9
Réunion 23
rhesus monkeys 67
rice 166–7
RNA 102
  interference 204–5
  micro 106, 124–5
  viral 126–7
roofing technology 148–9

satellite navigation 180
Saturn 265
schizophrenia 12
scratching *see* itching
screening for diseases 5–6
senile dementia *see* dementia
servers for websites 61–2
'shape memory' in metals 146–7
side effects of drugs 49–50
single nucleotide polymorphisms (SNPs) 194, 213–14
skin cancer 97–8

skin colour, link to environmental adaptation 95–6
sleep and sleep disorders 11–12, 43–4
sleep deprivation 75–6
sleep, effects on health 43–4, 75–76
sleep, effects on obesity 75–76
sleep patterns, infant, link to smoking 154–5
smell, sense of 136–7
smoking, effects on health 31–2, 37–8, 63–4, 154–5
solar technology 107–8
sound, effect on plants 166–7
Snuppy (cloned dog) 68
solar system, ancient 252–3
solar wind 253
songbirds 47–8, 217–18
space collisions 212
spawning of coral 101–2
speech, human 47–8
spiders 278
spinal cord 17
Spitzer Space Telescope 200, 271
squirrels 29–30
star 'mergers' 257–8
star systems, nearby 83–84
stem cell cloning 67–8
stem cells and stem cell therapy/transplants 35–6, 45–6, 67–8, 77–8, 223–4, 228–9, 244–5, 267
stomata 160–1, 189
strokes 140–1, 185
'suicide', cellular 93–4
sulforaphane 97
sunscreens and sunscreen technology 91–2, 98
sunspots, link to flooding 174–5
superbugs 94
suprachiasmatic nucleus 11
surgical techniques 146–7
swallowing mechanisms, in animals 156–7
syn-propanethial-S-oxide 15

tamoxifen 50
targeted gene therapy 18
targeted muscle re-innervation (TMR) 73–4
taste, sense of 128–9
Taxol 144–5
Teflon 103
telescopes, space 200, 231, 271
temperature sensitivity 249–50
terahertz radiation 58
testosterone 136, 261
'theory of mind' 30
thermometer, molecular 164–5
thrips 117–18
T lymphocytes 19
TMV (plant virus) 85
tobacco plants 53–4
toxins and antitoxins 93
traffic pollution 41–2
trainers, comfort vs price 80
transport 181–2
travel 180
transgenic crops 53–4
transplants *see* organ transplants

T-rays 57–8
trees 33–4, 117–18
triclosan 60
tumours and tumour formation 70, 208–9, 235–6
turtles 119–20
twins, identical 69–70

UV absorption by water 91–2
UV absorption in plants 91–2
UV radiation 97–8

vascular diseases and disorders 185–6, 278
vaso-constriction 21
vegetables 97–8
vehicles, alternative power for 232–3, 269–70
ventricular tachycardia 45
Verrucomicrobia 71
Viagra, natural 278
vibration as 'exercise', 99–100
vicrivoroc (drug) 178
virology 93–4, 263–4
virus, common cold 9
virus, flu 126–7
viruses and viral infections 17, 85–8, 105–6, 135, 178–9, 230, 263–4
visualisation techniques in pain control 56–7
vitamin C 31
volcanoes and volcanic eruptions, 23–4, 121, 239–40

'waggle dance' (bees) 61–2
Wangiella dermatitidis (fungus) 237–8
water, extraterrestrial 200–1
water, heavy metal contamination of 104
water pollution *see* pollution and pollutants
water, UV absorption by 92
water-saving strategies for crops 53–4
weather forecasting 39–40, 160–1, 174–5, 255–6
  extraterrestrial 271–2
websites 61–2
weight loss 99–100, 198–9
western scrub jays 29
white-crowned sparrows 217–18

x-rays 5–6, 28

yawning 221–2
yeast, use of to detect explosives 277
Yersinia pestis 263
YORP effect 153

zebra finches 47